LABORATORY ANIMAL HANDBOOKS NUM

Health and Safety in Laboratory Animal Facilities

Edited by

Margery Wood

Maurice W Smith

Published in the United Kingdom on behalf of Laboratory Animals Ltd (http://www.lal.org.uk) by The Royal Society of Medicine Press Limited, 1 Wimpole Street, London W1M 8AE.

British Library Cataloguing in Publication Data
A catalogue record for this book is available from the British Library
ISBN 1-85315-421-0

Typeset and printed in Great Britain

Contents

Contributors

Dr Michael J Dennis Centre for Applied Microbiology and Research, Porton Down, Salisbury, Wiltshire SP4 0JG

Mr Kevin Dolan Bioscientific Events, Petts Wood, Kent BR5 1EA

Dr Malcolm R Gamble Laboratory Animals Ltd, 20 Queensbury Place, London SW7 2DZ

Dr Susan Gordon Department of Occupational and Environmental Medicine, Imperial College School of Medicine at the National Heart and Lung Institute, Manresa Road, London SW3 6LR

Dr Ian Palotai SmithKline Beecham, New Frontiers Science Park, Third Avenue, Harlow, Essex CM19 5AW

Mr John Ryder Formerly of Pharmaco LSR, Eye, Suffolk IP23 7PX

Mr M W Smith M. S. Management Services, Cambridge, Cambs CB1 6HY

Professor David M Taylor Cardiff University, Department of Chemistry, PO Box 912, Cardiff, Glamorgan CF1 3TB

Dr Rosemary D Tee Department of Occupational and Environmental Medicine, Imperial College School of Medicine at the National Heart and Lung Institute, Manresa Road, London SW3 6LR

Dr Margery Wood CBD, Porton Down, Salisbury, Wiltshire SP4 0JQ

Preface

Since the second revised edition of *Safety in the animal house* was published as Laboratory Animal Handbooks 5 in 1981, there has been significant change in the design and function of animal houses. New concepts and techniques have been introduced into the care and use of laboratory animals and legal controls over both health and safety and animal welfare have developed further in the United Kingdom, the rest of the European Community and elsewhere. It is no longer valid to describe an animal house as a simple holding facility as it is often complex with modern equipment and many functions.

This new handbook, which replaces *Safety in the animal house*, takes account of current thinking and knowledge with new chapters on allergenic hazards and transgenic animals. The contributors show that health and safety provision is part of everyday working life and must be incorporated into working practices, both routine and specialized, and not treated as a separate issue. Guidance is given on assessing hazards and risks and how to eliminate or minimize them, but it is emphasized that each facility is unique and must be assessed locally by competent persons familiar with all aspects of the situation. Though our approach is orientated principally towards conditions prevailing in the United Kingdom, in our view it reflects best practice and the book is recommended to all those working with laboratory animals, regardless of their location.

We are extremely grateful to our splendidly patient and resilient authors: Malcolm Gamble, Susan Gordon & Rosemary Tee, Mike Dennis, Maurice Smith, Ian Palotai, David Taylor, John Ryder and Kevin Dolan, who have responded so well to the comments and requests of reviewers and editors.

The editors also thank the following colleagues who reviewed chapters or otherwise provided information or access to information to assist us in our task. They are: M Broster, M Burton, R Fosse, B Harvey, C Hetherington, S Leary, S Peel, R Soilleux and R

Wootton. In addition, our thanks are due to members of the Health and Safety Executive who have given a great deal of helpful advice and assistance during the preparation of this handbook.

However, in the end, the contents are the responsibility of the authors and editors who hope that this handbook proves to be a readable and up to date aid to understanding health and safety requirements in the laboratory animal facility.

Margery Wood **Maurice Smith**

Introduction to health and safety in laboratory animal facilities

Malcolm R Gamble

Contents

Foreword

Introduction

We are all vulnerable to danger, injury or loss, whether we are awake or asleep, and we accept this as part of the price to pay for living. By modifying our actions we can reduce these hazards and the risks of harm stemming from them, but to be completely free from them—the definition of safety—is an entirely different problem. As any golfer will tell you, although hazards are best avoided altogether, the trick is to identify each one before beginning the game.

This handbook is not a substitute for safety assessments or codes of practice, nor does it provide all the answers. However, safety awareness has often been called a state of mind and it is hoped that the specific areas covered here will enable relevant questions to be asked—even if particular answers cannot be found.

This chapter is an overview of hazards in general and is an introduction to the seven more detailed chapters that follow. All these chapters deal with the special nature of laboratory animal facilities and both widespread and particular hazards and risks to those who work in or visit them. We are all responsible, at some level, for the health and safety of ourselves and others.

Similarities between facilities

The last decade of this century has seen no slowing down in the diversity and development of laboratory animal facilities and the electronic complexity of their control systems. However, variations in size, construction, layout and function do not alter the common health and safety requirements that unite them all. The ethical containment of animals—together with their maintenance, use and welfare—should be at the heart of any animal house. This has to be placed alongside the increasing need for facility and personnel security. Compliance with national, regional or institutional legislation will affect the types of procedures undertaken and the management control of work routines, stocking densities and environmental conditions.

The management of staff safety encompasses establishment of management chains, suitable training and education, provision of information, risk assessment, flexible planning, monitoring and regular reviews of work practices and any protective equipment required. This has to be balanced with acceptable working conditions which will include occupational health provision, welfare, first aid and rest facilities (*see* Chapters 7 & 8 and National Research Council 1997).

Differences between facilities

The variable combination of hazards, however, leads to great diversity of potential risks, which will be influenced by individual persons, pieces of equipment, activities and types of animals involved. So each room should be considered separately as well as in relation to its contribution to the animal unit as a whole. All staff and visitors will add to the risks by their presence and the work they undertake, from animal technicians, scientists and veterinarians, to support staff, engineers, plant operators, drivers and laundry and stores staff. Types of activity will vary and could include risks from surgical procedures, toxic hazards, chemicals, infective agents and radiation. The size and temperament of the animals in use pose risks, as do their microbial and genetic status and allergenic potential for the staff in their vicinity.

Physical considerations

Building suitability

In the very first paper published in the first volume of the journal *Laboratory Animals*, in April 1967, Walker and Poppleton described the establishment of a specific pathogen-free (SPF) rat and mouse breeding unit behind a 'barrier' in a new building separated from their main laboratory. It was the need to standardize the microbiological conditions for the animals that required the organizational and physical restraints (barriers) to be developed to achieve this. However, nowadays, increasingly it is the need to maintain safe working practices in a healthy environment for both animals and staff that determines the design, fabric, contents and work routines of the animal house.

Housekeeping

Hazards associated with housekeeping can be avoided by commonsense precautions equally applicable to any domestic or work area. However, trips, slips and bumps account for a large proportion of recorded 'accidents' in animal houses. Water is probably the major hazard. Floors that are fogged, mopped or hosed should be made of an impermeable, non-slip material (whether wet or dry) that can be easily cleaned, and suitable non-slip footwear should be provided for staff. As a general rule water should not be left to aggregate in puddles; there should be sufficient fall in the floor surface to gullies or drains. Spillage of chemicals should be dealt with promptly and safely in an appropriate, predetermined way according to the manufacturer's/supplier's instructions. Disinfectant foot dips and damp pads at the entrances to barriered areas should always have slip hazard signs in prominent positions and be provided with a grab rail, as should rodent barriers within or at the entrance to buildings.

Regular floor cleaning is essential to contain both allergenic and potentially infective material and to remove items such as diet pellets which could cause slip hazards. Water is good for damping down dust and reducing inhalation hazards, but when used under pressure or carelessly could generate potentially harmful aerosols.

By their very nature the stores areas and corridors in animal houses have equipment, cages, diet and bedding continually passing through them. Care must be taken to avoid creating obstacles or trip hazards. Corridors must not be used as storage areas and should be wide enough to allow easy passage in each direction and to manoeuvre trolleys, racks, etc. Rodent barriers across doorways need to be effective and work routines should ensure their immediate replacement after access.

Stores, however large, nearly always become overused and materials have to be stacked to maximize the space available. Use of pallets and racking should be sensible and vigorously inspected by both management and other competent persons (e.g. Health and Safety Advisors). Use of powered forklift trucks or pallet trucks should also be specifically controlled and used only by properly trained staff of proven and certified competence. Compressed gas cylinders are best secured outside, under cover but well ventilated with appropriate warning signs posted. Gases may then be piped into the building through regulators that are serviced and tested appropriately.

The importance of personal hygiene cannot be over-emphasized and provision should be made for staff in the form of sufficient hand washbasins (with elbow taps), bactericidal soap dispensers, and showering facilities. Staff rest rooms and canteen facilities should be separate from the animal areas, and outer clothing, at least, should be changed when moving from one area to another.

Barriered animal facilities are at risk if staff do not comply with requirements for a complete change/shower, etc., on entry if laid down in local rules. Because of their isolated nature, changing areas should be provided with some means of summoning help (e.g. single press telephones, panic buttons or lone worker alarms) in an emergency. A name board or other device to indicate staff location as 'in' or 'out' can be checked at the beginning and end of the day or during an emergency roll call.

Plant and machinery

The major pieces of plant and machinery are usually situated in the service areas and are associated with the disposal of waste materials, cage cleaning, sterilization and heating and ventilation. The hazards associated with them are those involved with electricity, gas, steam, water, heat, fire and moving parts—individually and in combination. Safe working practices are enforced by various government standards, Regulations and Acts of Parliament in the UK and similar pieces of legislation in other countries, and are also being incorporated into several European Standards (*see* Chapters 7 and 8).

Regular maintenance, monitoring of performance and statutory testing of such plant are essential to ensure safety, effectiveness and efficiency. Records of all these activities should be kept. A good working relationship with maintenance and service engineers, with appropriate division of responsibilities, should be the rule.

Incinerators

Incinerators are a unique blend of hazards and have the potential to be the cause of injury, property damage and environmental pollution if mismanaged. Empty aerosol cans, organic solvents and scintillation fluid, quantities of paper and dense forms of plastic must be excluded from general or clinical waste and any sharps, e.g. scalpel blades or needles, should be in robust containers. The bags of waste should not be too heavy if manual handling is necessary and they should be waterproof and resist splitting when handled. Double-wrapping may be necessary where a risk of tearing or leaking is present or the outer surface of the inner bag is contaminated in some way. The mixture of items placed in the incinerator should be suitable for controlled and efficient burning. Monitoring panels should incorporate safety breakers and alarms to maintain safe conditions should the fuel supply or ignition malfunction or the temperature become excessive (to prevent damage to the chamber lining and exhaust flue stack) or waste incompletely burned. The controls should also allow for time locks on the charge door and burn-down times to ensure complete destruction of the load. If the incinerator is inside a building, care should be taken to ensure that the air pressures and supply volumes are balanced to maintain air movement over the load and to prevent blow-back into the operator space.

All incinerators in the UK burning clinical waste (which includes animal waste) must be authorized under the Environmental Protection Act (1990) and specifically its Prescribed Processes and Substances Regulations (1991) (*see* Chapters 7 and 8) and conform to specified plant standards. Local authorities control this licensing and carry out inspections to ensure that the monitoring of emissions, performance and control requirements are being carried out. An operational controller has to be identified who must ensure compliance with all standards. EC Directives are also in place to control waste disposal and the processing of animal waste (*see* Chapter 8).

Autoclaves

Autoclaves have increased in size and sophistication in recent years. Digital controls and printouts allow careful monitoring of performance, but additional, independent sterilization indicators should be incorporated into the load. Abort buttons, safety cut-outs and failure alarms are incorporated in the controls and pre-set cycles may be protected against inadvertent or ill-advised adjustment via password protection. However, care has to be taken in loading, both in terms of the materials themselves and how they are stacked to allow steam to permeate the load. Heat-proof hand and arm protection should be available for the removal of each load. Side panelling should always be locked as contact with the heated jacket or piping could cause burns. Autoclaves with walk-in chambers should have key-controlled door immobilizers and the operator should have unrestricted line of sight of the chamber when operating the door closure button. Doors should retract if an obstruction is met by the leading edge when closing. Regular certified safety pressure testing of the chamber as well as a rigorous maintenance schedule should be carried out and clear records kept.

Cagewashers

Cagewashers are generally well contained within a service area. With cabinet washers care should be taken with the positioning of pump and motors. They are often on top of the cabinet but if, for reasons of space, they need to be re-positioned, care should be taken to isolate and protect electrical terminals and supply wiring from becoming wet. Tunnel washers have low geared bar belts which must be guarded to prevent clothing becoming entangled. Emergency cut-off buttons should be placed near each end of the belt and not centrally or otherwise out of reach of staff loading or unloading the washer. Such machines may create an aerosol hazard as the washing chamber is not enclosed.

Heating and ventilation plant

Heating and ventilation equipment is generally contained within its own plant room and all servicing or inspections should be left to qualified engineers. Filtration, humidity control, heat exchanger, cooler, boiler and fan maintenance are skilled occupations. The animal house manager should have access to environmental records from plant or room detectors. It should be possible to detect when systems fail or go outside agreed limits via audible/visual alarms or computerized building management systems (BMS) which continually record and monitor the system.

Electricity

Perhaps the most widespread hazard is electricity. The use of residual current devices is strongly recommended as any problem can be readily detected and limited, with minimum damage. The Electricity at Work Regulations (1989) apply to animal houses, and competent persons should check correct functioning of all new electrical equipment (including portable appliances), particularly the types of fuses fitted and the safe connection to the supply circuit.

Switch socket outlets should have splash-proof covers in rooms or corridors which are periodically fogged, sprayed or washed down. Permanently plugged items should have leads held away from gnawing by animals, and temporarily plugged items should neither cause trip hazards by trailing across the floor nor be left lying in water. Refrigerators and freezers should be of a special, spark-proof kind if flammable solvents are to be stored in them or nearby or operations cause dust generation. Cabling and plugs should be checked periodically for wear or deterioration in keeping with current regulations.

Other equipment

Other hazards may be mechanical in nature, and relate to gaining access to dangerous parts of machinery (e.g. macerators, homogenizers, choppers, blades, etc). The likelihood of harm is low if the machinery is correctly operated with appropriate safeguards in place. In general, the production and maintenance of standard operating procedures for each piece of equipment is recommended and is a requirement in areas operating under good laboratory (GLP) or good manufacturing (GMP) practice.

Radiation hazards have great potential for harm, as exposure generally cannot be detected without the use of special equipment. The benefits and usefulness of electromagnetic radiation such as X-rays, ultraviolet (UV) light and microwaves need no explanation but their very ubiquitousness may lead to a casual attitude to them. Chapter 6 gives a comprehensive review of this topic.

Manual handling

Workplace injury records highlight the frequency of injuries caused by lifting, moving, pulling or pushing equipment. This is particularly relevant in the laboratory animal facility. It has been reported that nearly 50% of medical problems of animal workers are back strains (Aromaa 1989, quoted in Pelkonen 1994). Statutory requirements now control these operations in several countries (*see* Chapters 7 and 8).

A survey should be conducted of all such operations. Careful consideration should be given to handling animal cages, diet, bedding, trolleys, racks, as well as the animals themselves. The manual handling operation should be analysed taking into account body position, weight, size and shape of object, as well as the environment in which it takes place. The capability of the individual should also be borne in mind as well as their age, any disability and height. Certain staff may vary in their abilities (e.g. during pregnancy), and assistance or alternative staffing routines should be organized. Training courses in manual handling are now widely available.

Environment

This somewhat over-used term covers a wide variety of topics, many of which are dealt with in other chapters of this book. Factors affecting the working conditions of staff require the closest scrutiny. The Health and Safety at Work etc. Act (1974), subsequent workplace regulations in the UK and similar European workplace regulations cite the principal areas for concern. A few of the main areas will be considered here.

Lighting

There is a balance to be made between lighting levels bright enough to inspect the animals and carry out tasks but which will not damage the retinas of albino animals. It is known that staff will keep more brightly lit areas cleaner, and expert advice should be sought for optimum light levels and colour balance with surface finishes. Lights are generally best positioned behind sealed ceiling panels with access to them from above in the ceiling void or plant room. This reduces service access to the animal rooms.

Air supply

The levels of sophistication of heating, ventilation and air conditioning (HVAC) systems vary but they all must comply with local fire regulations. These can specify control features such as fail-safe dampers to isolate lengths of trunking, vents or meters in the event of smoke or fire. The design of room ventilation must not compromise the level of room containment provided by any fume cupboards, extract hoods or microbiological safety cabinets. Neither should it allow the dissemination of infectious organisms or toxic chemicals from one laboratory to another.

Directions and patterns of air flow can be used to support clean/dirty separation, and good air change rates with 100% fresh, conditioned air will aid the clearance of allergens, etc. Regular checking of pressure differences, supply and extract volumes and thermostat calibration could have far-reaching implications for both animals and staff in terms of health, safety and the scientific validity of the work in progress.

Noise

Noise is now recognized as a major hazard to staff in many areas of work and recreation. Irreversible damage can result from exposure to levels above 90 decibels (dBA) or even lower for extended periods of time. Discomfort alone may lead to carelessness and a tendency to rush jobs to escape from the noise.

Dogs go through understandable periods of agitation or excitement and vocalize at feeding times and when visitors enter their unit. Hence the housing of dogs can result in periods of noise exposure for staff caring for them. The installation of sound-absorbent material on walls, floors and ceilings can reduce the noise reverberation time but does little to reduce direct noise levels.

The mandatory wearing of sound-attenuating ear protectors will be required if noise assessments indicate that they are necessary; these assessments should only be done by designated competent persons. In this and other instances (e.g. wearing an air-feed helmet) where hearing ability is impaired, care should be taken that visual fire or intruder alarms are installed, as well as other communication aids to indicate telephones and door bells, suitably positioned so that the lights can be easily seen during work routines.

Fire

Of all the general hazards summarized here, it is fire that has the most serious, far-reaching and immediate consequences. Fire is frequently associated with electricity, poor housekeeping and chemicals. Formal preventive measures and inspections are vital and should include provision, maintenance and monitoring of signs, exit routes, fire doors, fire-fighting equipment, sensors and alarms.

Close liaison with the local fire authority is necessary, particularly during the design and construction of new premises, to ensure compliance with local regulations, provision of appropriate signs and correct numbers and types of fire extinguishers. A record should be kept of fire extinguisher inspection and maintenance.

Proper housekeeping should include regular inspection of all electrical appliances and replacement of worn or ageing flex, provision of proper containment for flammable chemicals, spark-free fridges and a clean and uncluttered environment. Care should be taken that any automated air conditioning systems or electrically interlocked doors that 'fail-safe' on the activation of the fire alarm correctly revert to normal once the systems have been re-set.

Biological considerations

Laboratory animal allergy

Laboratory animal allergy (LAA) or allergy to laboratory animals (ALA) is a very important and widespread occupational health problem and its causes and management are detailed in Chapter 2. This hazard is unconnected with the health status of animals. Surveys have indicated that of the 30,000 people working with laboratory animals in the UK, 15–35% may develop animal-related allergies (Health and Safety Commission 1990). It is not possible to predict who will develop symptoms, which may vary from mild nasal discharge or sneezing to conjunctivitis, urticaria and asthma.

Allergens are 'substances hazardous to health' and as such should be included in COSHH (Control of Substances Hazardous to Health) assessments. The greatest number of allergy cases are associated with rodents, possibly because far more of these are involved in scientific procedures than other species. A clear policy should be devised in all areas using mammals, insects and birds as to the actions necessary to control the exposure of staff via the respiratory tract and skin contact. Design of laboratory animal units and work routines adopted and enforced in them are two major areas where control is possible. The two fundamental principles are to minimize the amount of allergenic material liberated into the air and to remove what is present as effectively as possible.

Urine, hair, fur, dander, body tissues and secretions may all contain varying amounts of allergenic material depending on the age, sex, strain and species of animals in use. Well designed HVAC systems with 15–20 air changes per hour of conditioned, filtered air distributed efficiently within the animal room are essential. The primary air pattern and direction within the operator space is also of direct relevance. Impermeable surfaces that can be cleaned easily and do not trap dust are essential to aid control.

Even if a significant air dilution of allergens occurs there is still a need to control the spread away from the animal area through doors and along corridors. This can be achieved by effective scavenging of air and differential air pressures creating a clean to dirty flow. Installation of electrostatic precipitators to scrub the air is a further option.

Infection

All conventionally housed laboratory animals naturally carry a wide range of microorganisms internally and on the surface of their bodies. Many organisms are harmless and can be matched by those carried by staff working with the animals. However, some organisms are pathogens and may give rise to infections if they gain access to a new, susceptible host. They may also mutate in a new host species into a more harmful form. Infection can occur via the skin, alimentary tract, respiratory tract, conjunctiva and genitourinary system, depending on the microorganisms involved. Tetanus immunization of staff should be standard practice.

The best way of assessing risks to staff is to identify which pathogenic organisms are present. Confirming the absence of known pathogens in a single sample is generally not sufficient. Animals, bred or bought, should be screened at regular intervals and sentinel animals (microbiologically 'clean' and especially susceptible) should be housed alongside experimental animals as potential early warning indicators (amplifiers) of otherwise subclinical infection. Long-term experiments should also include additional, untreated animals for screening at the end of the test to confirm sentinel results. Animals of different microbial status should be housed separately and appropriate measures taken to limit spread of infection to other animals and staff.

Regular occupational health monitoring of all staff working with laboratory animals is essential where an infection hazard exists or is suspected. This and other major areas for consideration are detailed in Chapter 3, where the experimental use of microorganisms is also considered. It must be noted that, compared to their normal counterparts, some genetically modified animals (*see* Chapter 4) may be of an inferior microbial status and may, because of their altered genetic makeup, be more susceptible to a greater variety of microorganisms. Hence they may be a greater risk to staff.

Chemical considerations

Disinfectants, cleaning and sterilizing agents

Anyone who has had to close down an animal-based laboratory unit or move one to another site will have needed only to open cupboards or clear store areas to realize the wide range of chemical hazards that are or were in regular use. Acids, alkalis, degreasers, descalers, detergents, bleach and other disinfectants, dyes, stains, drugs and several unidentifiable bottles and containers are commonplace. Many are irritant, corrosive, toxic and potentially carcinogenic or teratogenic.

Identifying the risks and taking steps to reduce them are detailed in Chapters 5 and 7. The COSHH Regulations (1988–1999) have enforced assessments of all such materials and safety data sheets are now widely available. Health surveillance is appropriate for all staff exposed to known or suspected carcinogenic substances (novel pharmaceutical compounds should be treated as potential carcinogens until their safety has been proven). It is only by regular safety inspections, rigorous housekeeping and adherence to standard operating procedures (SOPs) and handling instructions that the risks can be reduced.

Anaesthetic gases

The use of inhalation anaesthetics poses additional problems to those who use them. Although the use of scavenging devices is increasing, there are still many laboratories and surgeries which have inadequate ventilation. Exposure to some anaesthetic agents may cause malignancies, as detailed by Green (1979).

Bites and scratches

Careful and correct handling of all species of animals will reduce the incidence of bites, scratches and other injuries to a minimum. Such injuries are usually of a minor nature, but should be referred to the occupational health service for treatment and monitoring where a risk of infection or contamination by chemicals or radioactive materials is possible. If venomous animals are held, appropriate antidotes should be readily available.

The use of microbiologically defined animals has reduced the infection hazard, as has the reduction in numbers of imported non-human primates in use. Approved protective clothing should always be worn when handling any primates, including masks, gowns, gloves, visors and overshoes when appropriate. All injuries to staff, however minor, should be referred immediately to a local first-aider and the incident reported.

Personal protective equipment (PPE)

The PPE at Work Regulations have controlled the use of such equipment in the UK since 1992. The principle put forward in the Health and Safety at Work etc. Act (1974) was that PPE should be the last resort. Methods of working and engineering controls should be devised to eliminate or reduce the need for operator personal protection. Assessment on a continual basis of how the job is carried out is vital for this to succeed. Completing or updating the assessments is less complex than deciding which type or make of equipment is most suitable. In the UK currently there are many companies specializing in the manufacture and sale of PPE.

Providing PPE is no guarantee that it will be used correctly or maintained in a suitable condition. Training has to be given and several suppliers provide this as part of their sales service. It must be realized that it is not unusual for PPE to have a decremental effect on physical activity. For example, thick gloves may reduce dexterity and sensitivity and respirators may impair vision and be tiring to wear. Hence staff must understand why equipment is needed and the consequences of non-compliance. PPE can protect from physical, biological and chemical hazards but should be distinguished from non-critical protection afforded by items of clothing or uniform adopted for reasons other than health and safety.

Waste disposal

Society is recognizing that it has a responsibility for the environment and companies and institutes are usually involved in formulating policies to improve their performance in this area. Waste minimization is to be encouraged, as is the recycling of materials. Due regard should also be taken of the effects on the environment of any wastes produced. The identification and control of hazards associated with these functions are therefore of great importance.

Waste materials should be classified according to their constituents and their particular hazards. Policies should then be formulated for their controlled removal.

Disposal of chemical and radioactive waste will be dealt with in their separate chapters. Animal carcasses, soiled bedding and food waste, infected material and sharps all pose a direct hazard to those handling them. Destruction in-house reduces the amount of handling and limits the involvement of others, but this may not be possible. Clear instructions and equipment (e.g. stout polythene sacks, sharps containers, etc.) must be provided to contain waste from source to final product, be it incinerator ash, macerator discharge or council landfill sites. Responsibility for safe disposal of workplace waste does not end with handing it over to a contractor.

Security

Laboratory animal facilities and the people who work in them are becoming increasingly subjected to various levels of threat, aggression and violence by small groups of individuals so fanatical in their belief in the single issue of animal rights that they are prepared to cause criminal damage and assault in support of their cause. Increasing resources have to be deployed so that buildings and people can be protected from damage. Procedures should be in place for detecting and responding to any incidents as they occur.

Staff working alone in laboratory or animal areas pose a particular problem of supervision and protection from outside influences as well as accidents or illness. Any person out of line of sight of anyone else for any prolonged period (say 30 minutes) can be better protected by means of a lone-worker alarm. The use of these devices, which are radio alarm transmitters having both panic buttons and 'lack of movement' detectors, allows constant monitoring by security staff.

References

Green NG (1979) Euthanasia. In: *Animal Anaesthesia* (Green CJ, ed.). Laboratory Animal Handbooks No. 8. London: Laboratory Animals Ltd, pp 237–41

Health And Safety Commission (1990) *What you should know about allergy to laboratory animals*. Education Services Advisory Committee. Sudbury: HSE Books

National Research Council (1997) *Occupational Health and Safety in the Care and Use of Research Animals*. Washington: National Academic Press

Pelkonen KHO (1994) Health and safety in work with laboratory animals. In: *Welfare and Science* (Bunyan J, ed). Proceedings of the 5th FELASA Symposium, Brighton, 1993. London: Royal Society of Medicine Press, pp 141–4

Walker AIT, Poppleton WRA (1967) The establishment of a specific-pathogen-free (SPF) rat and mouse breeding unit. *Laboratory Animals* **1**, 1–5

2

Allergenic hazards

Susan Gordon and Rosemary D Tee

Contents

Introduction

Laboratory animal allergy (LAA) is the most widespread occupational health problem of those exposed to these animals. Exposure occurs in the pharmaceutical industry, in universities and research institutes in those involved in animal care, research and associated occupations. The most common symptoms are rhinitis, conjunctivitis, contact urticaria and asthma, which develop following inhalation of, and other contact with, animal excreta and secretions. Many individuals who develop LAA and nearly all those with asthma have IgE antibody specific for laboratory animal derived allergens.

Since January 1989, British chest physicians and occupational physicians have reported new cases of lung disease to the Surveillance of Work Related and Occupational Respiratory Disease (SWORD) project (Meredith *et al.* 1991). In these reports, those exposed to laboratory animals had an incidence rate of asthma of 19 per 100,000 per year in the two years 1989/1990 compared with a mean rate of only 0.22 in the general working community. Laboratory animals remain one of the three most commonly reported agents causing occupational asthma, and the number of cases reported is apparently rising (Ross *et al.* 1995).

As the UK Home Office now publishes only the number of animal experiments performed annually, it is even more difficult to estimate the number of people exposed (Home Office 1987), but in 1980 it was estimated that 32,000 persons worked with laboratory animals in the United Kingdom (Cockcroft *et al.* 1981). Cross-sectional surveys of laboratory animal workers have consistently found prevalence rates of LAA between 17% and 44% (Agrup *et al.* 1986, Cockcroft *et al.* 1981, Schumacher *et al.* 1981, Slovak & Hill 1981, Venables *et al.* 1988). It is clear from the continuing high prevalence of LAA worldwide that individuals with animal contact remain exposed to sufficient allergen to become sensitized and develop symptoms (Aoyama *et al.* 1992, Bryant *et al.* 1995, Fuortes *et al.* 1996).

It has been known for some time that extracts of urine proteins of rats and mice elicit immediate skin-prick test responses and, when inhaled, provoke asthmatic reactions in sensitized workers (Newman Taylor *et al.* 1977). The majority of cases of LAA are attributed to rats and mice, and detailed information on the possible allergen sources for these species will be discussed. Allergic symptoms also occur amongst those working with rabbits, guineapigs and insects, such as locusts, and the pertinent allergen sources for these will be considered. Brief reference will be made to allergy associated with chemicals.

In the UK following the introduction of the Health and Safety at Work etc. Act of 1974, in the event of a person becoming ill at a work place, the employer is responsible and liable for redress in the event of negligence. Occupational allergy to laboratory animals is within the scope of the Control of Substances Hazardous to Health Regulations of 1999 (Health and Safety Commission 1999). In addition, occupational asthma caused by laboratory animals must be notified to the Health and Safety Executive under the Reporting of Injuries, Diseases and Dangerous Occurrences Regulations (RIDDOR) of 1995. Occupational asthma is a prescribed occupational disease and as such qualifies for state disability benefit.

Definitions of terms used in the text

Allergen. A substance that elicits a specific IgE antibody response.

Allergy is a synonym for hypersensitivity which implies a heightened reactivity to antigen. The term is used especially for immediate hypersensitivity which is antibody-mediated.

Anaphylaxis. An acute immediate hypersensitivity following exposure of a primed subject to allergen. Anaphylaxis may be generalized (anaphylactic shock) or localized to the site of entry.

Antibody is a protein produced by the immune system with the molecular properties of an immunoglobulin and which is capable of specific combination with antigen.

Antigen. A substance that elicits a specific immune response when introduced into the body. This takes the form either of antibody production or cell-mediated immunity, or of specific immunological tolerance.

Asthma is airway narrowing which reverses over short periods of time, either spontaneously or as a result of drug treatment.

Atopy. Persons who have a hereditary predisposition to producing IgE antibody to inhalant allergens are said to be atopic. Atopy may be defined clinically by family or personal history, or by a positive skin-prick test reaction to one or more common allergens, e.g. house dust mite (*Dermatophagoides pteronyssinus*), grass pollen or cat fur.

CIE. Crossed immunoelectrophoresis is a technique where a crude extract containing antigens is first separated by electrophoresis in one direction and then electrophoresed again at right angles into rabbit antibodies specific to the extract. Precipitin peaks develop and represent the antigens present in the extract as recognized by rabbit antibodies.

CRIE. Crossed radioimmunoelectrophoresis identifies specific IgE antibodies in patients' sera that bind to individual allergens. Following CIE, the allergen-antibody complexes are visualized on X-ray film after autoradiography (sequential incubation with patients' sera and ^{125}I anti-IgE). Major allergens may be defined as ones that bind IgE from $\geq$50% of sera with a given intensity of radiostain.

Immunoglobulin E (IgE). The immunoglobulin associated with reagin or homocytotropic antibody activity in man. The antibody fixes to tissue cells of the same species so that on reaction with antigen, histamine and other vasoactive agents are released. It is present in serum in very low concentration (20–500 ng/ml) but elevated in allergic asthma and rhinitis and in intestinal helminth infections.

RAST. The radioallergosorbent technique is an *in vitro* radioimmunoassay for the measurement of allergen-specific IgE antibodies.

RAST inhibition. A modification of RAST used in the standardization of allergen extracts and in the measurement of aeroallergen concentration.

RUA (rat urinary aeroallergen) is the concentration of rat urinary allergen measured in the atmosphere by means of immunoassay of air sample filtrates.

Skin-prick test. The detection of allergy to specific allergens through the production of a wheal and flare response by pricking the skin through droplets of allergen extract.

Occupational asthma is asthma induced by a sensitizing agent inhaled at work.

Peak flow measurements. The subject makes and records repeated measurements of peak expiratory flow rate (PEFR) from waking to sleeping using a peak flow meter. This should be done for a month to cover periods both at work and a week of holiday to see if there is deterioration in lung function during work periods and improvement away from work.

Epidemiological terms

Cohort is any designated group of persons who are followed or traced over a period of time.

Cross-sectional study or prevalence study is one that examines the relationship between the disease (or other health-related characteristics) and other variables of interest as they exist in a defined population at one particular time.

Incidence rate is the number of new cases of a disease occurring in the population during a specified period of time, expressed as a percentage of the number of persons exposed to risk of developing the disease during that period of time.

Longitudinal study or cohort study. Observation of a population for a sufficient number of years to generate reliable incidence rates of disease (or other outcome) in relation to a certain exposure.

Prevalence rate is the current number of cases of disease present in the population at a specified time, expressed as a percentage of the number of persons in the population at that specified time.

Epidemiology

There have been several cross-sectional studies reported of workforces exposed to laboratory animals (Agrup *et al.* 1986, Beeson *et al.* 1983, Cockcroft *et al.* 1981, Lincoln *et al.* 1974, Schumacher *et al.* 1981, Slovak & Hill 1981, Venables *et al.* 1988). These have examined allergic disease prevalence in current employees, the specific IgE antibody response and the factors that may predispose to LAA. The results of these studies have been very consistent, the prevalence of LAA being about 30%. Approximately 10% of the exposed population (approximately one-third of those with LAA) may develop asthma. Symptoms may occur between one month and several years after first exposure to animals, with a mean interval of some 2–3 years. The most common symptoms reported were rhinitis and conjunctivitis (occurring in most subjects) with contact urticaria occurring in 4–14% of allergic subjects where it may be the only manifestation of LAA. Asthma usually developed after the onset of other allergic symptoms and was associated with the presence of specific IgE. This antibody response was not necessarily found with other related symptoms where up to 50% of subjects had no detectable IgE. Asthma occurred some five times more frequently in atopic workers but they had no increased risk of developing rhinitis and urticaria.

Workers with LAA may develop multiple sensitivities to different laboratory animals. Rats and mice are the primary sensitivity in 60–70% of cases, but this may be a reflection of these being the most commonly used species. Guineapigs are implicated in 30–40% of cases and rabbits in up to 70%. This high figure for rabbits may be confounded by some symptoms being of an irritant, rather than allergic, nature (Aoyama *et al.* 1992).

A major problem of cross-sectional studies is that they examine a 'survivor' population in that those who develop disease, particularly asthma, are likely to avoid exposure to its cause. Two studies have reported an inverse relationship between the prevalence of chest symptoms and the duration of employment, suggesting that those with chest symptoms subsequently avoid exposure to animals or leave the employment (Kibby *et al.* 1989, Venables *et al.* 1988). There are no documented cases of a natural resolution of symptoms in those with LAA who continue to be exposed.

Cohort or longitudinal studies, where workers including those who have left employment are studied over several years, provide information on incident cases. Two such studies have been reported, but neither followed up the 'leavers'. Kibby *et al.* (1989) studied an American governmental research institute which had an incidence rate of 13% over the two years of the study. A more detailed seven-year study of a British pharmaceutical company reported that the incidence rate in annually recruited new workers fell by approximately two-thirds (mean of 34% in the first three years to a mean of 12% in the last three years), probably due to the improved work practices on the site and the use of simple masks (Botham *et al.* 1987, Botham & Teasdale 1987). Atopic workers were seven times more likely than non-atopics to develop LAA in the first year. However, when the first three years were analysed together, the increased risk among atopics was only two-fold more than in non-atopics, implying that the primary effect of atopy was to cause the symptoms of LAA to occur earlier.

Insects have also been shown to be potent sensitizers. It is among workers with insects that the majority of work-related allergies have been diagnosed. These would include carers, entomologists, toxicologists, geneticists and biochemists. In a review, Bellas (1990) reported that an average of 30% of those exposed to insects became sensitized to them.

Moths are a serious problem to those who come into intimate contact with the adults (Bauer & Patnode 1984) and several species of *Diptera* (flies) have been reported to cause allergy. These include *Drosophila* (Spieksma *et al.* 1986) and the screw-worm in the

United States of America (Wirtz 1980), house flies in the United Kingdom (Tee *et al.* 1985) and *Chironomids* in Germany (Baur 1982). Various species of grasshopper (Soparkar *et al.* 1993) and locust have proved troublesome in several studies around the world. One study of 35 research centre workers reported a high prevalence of allergic symptoms (Tee *et al.* 1988). Of 15 currently exposed to locusts, five had rhinitis, contact urticaria and asthma, three had rhinitis and urticaria and a further one had rhinitis only. A positive skin-prick test to locust extracts was found in all the nine symptomatic workers but in only three of 26 workers without symptoms. Atopy was not an enhancing factor for allergy to locusts. Other insects which have been identified as a cause of occupational allergy and have promoted specific IgE responses among exposed workers include cockroach (*Blatella* and *Periplaneta* species, Chapman 1993) and stick insects (Perlman 1958).

Diagnosis and pathogenesis of LAA

Diagnosis of LAA is mainly clinically based on the connection between a good history of a suitable animal exposure and allergic symptoms. Eye and nose symptoms are characterized by rhino-conjunctivitis with nasal watery discharge and much irritation. Asthma is the main chest symptom with periodic wheezing, cough and dyspnoea. Skin rashes are either general or contact urticaria (Agrup & Sjöstedt 1985). In the latter the wheals occur only on the contact area, usually on the face, hands and underarms (Rudzki *et al.* 1981). Occasionally, angio-oedema occurs but anaphylactic reactions with severe systemic effects are extremely rare (Teasdale *et al.* 1993, Watt & McSharry 1996). Once a person is sensitized, symptoms such as urticaria and rhinoconjunctivits develop very shortly after re-exposure, usually within 10 minutes; asthma may develop more slowly. Hunskaar and Fosse (1990) pooled data from 13 prevalence studies and found that the typical prevalence rate of symptoms for every 10 people with LAA were: eight had rhinoconjunctivitis, four had asthma and four had urticaria. A progressive pattern of symptom severity is seen in many patients. Initial contact with animals induces rhinitis often with conjunctivitis (Agrup *et al.* 1986, Davies *et al.* 1983a, Lutsky 1980). Asthma appears to be an end-stage of LAA but only a proportion of those who develop rhinitis progress to asthma; asthma in the absence of rhinoconjunctivitis is unusual.

Confirmation of the diagnosis of LAA, in the context of the subject's case history, can be obtained by a positive specific IgE response. Work-related asthma can also be identified by serial measurements of peak flow rate with deterioration in peak flow during periods at work and improvement during absences from work (Newman Taylor & Gordon 1993). Inhalation testing with specific animal extracts is now rarely necessary. Almost all asthmatic workers react positively to both specific skin-prick tests and the *in vitro* RAST, and there is a good correlation of results obtained from these two tests (Davies *et al.* 1983a). Only 50% of those with other symptoms of LAA have specific IgE, so these tests are not useful in those without asthma.

The increased risk of atopic individuals developing asthma due to laboratory animal allergens has been discussed. The risk of rhinitis and urticaria is no greater among atopics than non-atopics. Atopy should probably not be regarded as a bar to employment.

Allergenic hazards and their sources

Recent studies indicate that the majority of clinically important aeroallergens are biochemically active. A wide range of properties has been demonstrated but most possess either enzymatic acitivity, enzyme inhibitory activity, have regulatory properties or are transport proteins, e.g. Rat n 1 and Mus m 1 (Stewart & Thompson 1996).

Rat (Rattus norvegicus)

In 1977 the urine of rats and mice was identified as an important source of allergen (Newman Taylor *et al.* 1977). The low molecular weight (MW) constituents of the urine of adult male animals were particularly potent at causing bronchoconstriction and positive skin-prick test responses in sensitive subjects. Detailed biochemical analysis has identified the most important, or 'major', allergens as glycoproteins with the electrophoretic mobility of a prealbumin (21 kDa, also called *Rat n*1A, Rat n 2 but now termed Rat n 1.01) and α_{2u}-globulin (17 kDa, also called *Rat n*1B, Rat n 1 but now termed Rat n 1.02) (Longbottom 1980, 1983, Gordon *et al.* 1993a). The amino acid sequence of these allergens has been determined. They have been shown to constitute different forms of the same parent protein and have been identified as members of the lipocalin family of proteins (Bayard *et al.* 1996). A third 'major' allergen has recently been identified with a molecular weight of 23 kDa (Gordon *et al.* 1993a). In this study, all of the 83 rat urine-sensitive subjects had IgE to one or more of these low molecular weight proteins and five additional allergens were present with molecular weights of 44, 51, 63, 68 and 75 kDa.

The composition of rat urine is dependent on the age and sex of the animal. Prealbumin (Rat n 1.01) is present in the urine of both male and female adult rats (Longbottom 1983, Gordon *et al.* 1993a). There is no significant difference in the urinary allergens from different strains of rat (Lutsky *et al.* 1985). Alpha$_{2u}$-globulin (Rat n 1.02) is synthesized in the liver under complex hormonal control (Bond 1962, Roy *et al.* 1983) and has been shown to have important pheromonal properties (Bacchini *et al.* 1992). Testosterone is essential for the process and hence significant levels of α_{2u}-globulin are produced by adult male rats (Roy & Neuhaus 1967). Alpha$_{2u}$-globulin, secreted via the kidneys, may be detected in the urine of 40-day-old male rats (Roy & Neuhaus 1967) and reaches adult levels at puberty (Gordon *et al.* 1993a). The presence of α_{2u}-globulin in female rat urine is controversial, although two recent reports have identified proteins with apparent immunological similarity to the male protein, although present in lower amounts (Vandoren *et al.* 1983, Gordon *et al.* 1993a). The urine, or contaminated material such as soiled litter, from all rats should therefore be considered a potent source of rat allergens. Rat faeces are not thought to contain significant amounts of allergen (Walls & Longbottom 1983a, b).

All rat strains develop chronic renal disease spontaneously (Burek *et al.* 1988) due directly to the production of α_{2u}-globulin and its ability to bind toxic chemicals (Borghoff *et al.* 1990). The disease affects all male rats, but the incidence and severity is less in female rats (Burek *et al.* 1988). Male rats therefore secrete increasing levels of serum proteins in their urine as they mature (Alt *et al.* 1980, Burek *et al.* 1988, Dinh *et al.* 1965, Roy & Neuhaus 1966). Sprague-Dawley rats develop chronic progressive nephropathy and resultant proteinuria at approximately twice the rate of Wistar rats (Weaver *et al.* 1975). Albumin (molecular weight 68 kDa) and transferrin (molecular weight 75 kDa) are allergenic for nearly 30% of rat-sensitive subjects (Gordon *et al.* 1993a). The handling of urine from ageing rats, in addition to the dissection or handling of rat tissue, may therefore result in exposure to rat serum proteins.

Other sources of rat allergen which have been studied in less detail are saliva (Walls & Longbottom 1983a, b), and the pelt or fur of the animals (Ohman *et al.* 1975, Walls & Longbottom 1983a, b). Proteins with sequence homology to a_{2u}-globulin have been additionally identified in the submaxillary glands (Laperche *et al.* 1983) and modified sebaceous glands (Mancini *et al.* 1989) of the rat. A preliminary report of the allergens in male rat hair using sera from 77 subjects and immunoblotting techniques identified 20 allergens with five 'major' allergens at 55, 51, 19, 17 kDa and a very high molecular weight material (Gordon *et al.* 1993b). A large amount of allergenic cross-reactivity was noted between rat urine and hair.

In summary, the most important sources of rat allergens are urine, pelt and serum. Exposure to these may occur through the direct handling of animals or via soiled litter. Due to the potency of the airborne allergens, exposure may also occur via the handling of soiled clothing or equipment, or even entering areas where rats have recently been.

Mouse (Mus musculus)

The main component of mouse urine prealbumin (also called *Mus m*1 and Major Urinary Protein complex MUP) is a 17.5 kDa protein of hepatic origin which is secreted in the urine in an analogous manner to α_{2u}-globulin in the rat (Finlayson *et al.* 1965, Longbottom 1980, Lorusso *et al.* 1986, Reuter *et al.* 1968, Sirganian & Sandberg 1979). The speculation that mouse urine has pheromonal properties (Lombardi *et al.* 1976, Shaw *et al.* 1983) has recently been confirmed, and the two pheromones that are bound are common to rats and mice (Bacchini *et al.* 1992). Proteins with close sequence homology to prealbumin have also been found in several secretory tissues such as the submaxillary, lachrymal and mammary glands (Shaw *et al.* 1983). Antigenic similarity of salivary gland extracts to urinary prealbumin has been found by Finlayson *et al.* (1965) but not by Price & Longbottom (1987). Two major allergens have been identified: urinary prealbumin (Ag 1 or Mus m 1) and a fur-derived allergen (Ag 3, Mus m 2) (Price & Longbottom 1987). The fur-derived allergen is a protein with a similar molecular weight to the urinary prealbumin (16 kDa) but which contains polysaccharide residues. It is secreted from the hair follicle and coats the stratum corneum and the hair shafts (Longbottom & Price 1987, Price & Longbottom 1989). Albumin present in the pelt, epidermis and serum has been identified as an allergen in some sensitive individuals (Ohman *et al.* 1975, Schumacher 1980, Sirganian & Sandberg 1979).

Minor differences in the urinary proteins from different strains of mouse have been noted (Finlayson *et al.* 1974), but these do not affect the allergenicity (Schumacher *et al.* 1981). There is a high degree of homology between rat α_{2u}-globulin and mouse prealbumin (Clark *et al.* 1984, Hastie *et al.* 1979). The cross-reactivity between the two proteins has been confirmed by the ability of purified α_{2u}-globulin to cause proliferation of MUP-reactive clones (Gurka *et al.*

1989) and in RAST inhibition experiments (Gordon *et al.* 1994b, Longbottom 1983).

In summary, the most important sources of mouse allergen are the urine and fur, and the handling of animals or soiled litter poses an allergenic hazard.

Guineapig (Cavia porcellus)

Skin-prick testing and RAST inhibition studies have confirmed that, in order of importance, fur, saliva, and urine are allergenic for most guineapig-sensitized subjects. Approximately half of the subjects had IgE specific for guineapig serum and albumin. Many proteins have been identified in guineapig pelt by electrophoretic techniques (Ohman *et al.* 1975, Walls *et al.* 1985 a, b). Swanson and co-workers (1984) showed that the major allergens in pelt had a MW of 10–22 kDa, and that guineapig urine contains many acidic proteins, the most allergenic of which had MWs of 75–50 kDa. CRIE studies have identified 14 allergens in guineapig-associated dust, four of which were 'major' allergens and all of which were present in extracts of guineapig dander, fur, saliva and urine (Walls *et al.* 1985 a, b). Guineapig faeces contain little allergen.

The allergenic hazards in handling guineapigs are the same as for rats and mice, with the exception that the fur of guineapigs is the most important source of allergen in this species.

Rabbit (Oryctolagus cuniculus)

The urine, fur, saliva and dander/pelt have all been found to contain rabbit allergens (Ohman *et al.* 1975, Price & Longbottom 1986, 1988a, Warner & Longbottom 1991). The major rabbit allergen (Ag R1) has been identified by CIE and shown to be a glycoprotein with a MW of 17 kDa. Saliva and fur are the most potent source of Ag R1. Albumin and immunoglobulin light chain dimers have been identified as minor allergens in pelt, serum and urine respectively.

Other mammals

Allergy to cats and dogs may be present in up to 30% of the general population. The allergens produced by cats (*Felis domesticus*) and dogs (*Canis familiaris*) have been reviewed recently by Schou (1993). Important sources of allergen in both species are the pelt, dander, saliva and urine. The 'major' allergens (Fel d 1 and Can f 1) are probably of secretory origin and are present in high concentrations in pelt and saliva. Interestingly, the α-chain of the dimeric Fel d 1 molecule shares 28% homology in its amino acid sequence with rabbit uteroglobulin (Morgenstern *et al.* 1991). The serum protein albumin, also found in pelt, is probably a minor allergen but may contribute to allergenic cross-reactivity observed between cats and dogs. There is little evidence for breed-specific allergens although properties of the pelt, such as the length of hair, may influence allergen dispersion.

There have been few descriptions of allergy to gerbils (McGivern *et al.* 1985), monkeys (Petry *et al.* 1985) and marmosets (A J Newman Taylor, personal communication), and occupational allergy to these species is, presumably, rare. No detailed studies of the allergens involved have been undertaken, although skin test (scratch method) reactivity to monkey albumin has been reported among those sensitive to other mammalian albumins (Simon 1941).

Insects

Several species of insect have been linked to occupational allergy and, in some cases, detailed investigation of the allergens have been undertaken. The sources of the allergen can be faeces, secretions, hairs, setae, scales and dried exuviae (discarded exoskeletons). Tee and co-workers (1985) in their study of the housefly (*Musca domestica*) demonstrated that the airborne fly allergen(s) originated in part from the dust collecting in the fly-rearing cage and were probably of faecal origin. Similarly, faeces and the peritrophic membrane are the most likely source of airborne locust allergens in the species *Schistocerca gregaria* and *Locusta migratoria* (Tee *et al.* 1988).

Other allergenic hazards

Fresh bedding and food have been shown to be non-allergenic for those sensitive to rats and guineapigs (Walls & Longbottom 1983b, Walls *et al.* 1985b). However, stored foodstuffs and bedding may be contaminated with storage mites. Storage mites such as *Lepidoglyphus destructor*, like the house dust mite (*Dermatophagoides pteronyssinus*), are present widely in many environments and are a significant cause of occupational allergy amongst farmers (van Hage-Hamsten 1992) and a cause of a high sensitization rate in bakers (Tee *et al.* 1992).

Chemicals such as formaldehyde (Burge *et al.* 1985) and glutaraldehyde (Cullinan *et al.* 1992) may be associated with respiratory symptoms in occupational groups such as nurses and radiographers. The pathogenicity of such cases is complex and may be at least partly attributable to their irritant nature. There has been only one report identifying formaldehyde-specific IgE in the serum of a subject with formaldehyde-induced bronchospasm (Grammer *et al.* 1993).

Contact urticaria and occasionally rhinoconjunctivitis and asthma may be caused by latex gloves (for review see Levy *et al.* 1992). All those who regularly wear latex gloves are at risk, although the prevalence rate of latex specific IgE is approximately 5% (Committee Report 1993). 'Hypo-allergenic' gloves are available which may be more readily tolerated (Cormio *et al.* 1993). It is now possible to quantify airborne latex allergen, and recent reports have shown how both the number and type of gloves (e.g. powdered and non-powdered, high and low allergen) may influence aeroallergen levels (Heilman *et al.* 1996, Swanson *et al.* 1994).

Although the allergenic sources discussed in this section are proven causes of asthma and rhinoconjunctivitis in the occupational settings specified, their relevance to those employed primarily with animals is unknown. The greatest risk of sensitization for animal workers remains the animals themselves.

Measurement of airborne allergens in animal facilities

Methods of measurement

The quantification of airborne allergenic matter was restricted initially to that with a distinct morphological appearance (e.g. pollen grains or mite bodies), or which could be cultured (mould spores) or chemically assayed (e.g. isocyanates) (Newman Taylor & Tee 1989). The measurement of amorphous allergen, such as animal allergens, became possible with the development of immunoassays.

Several different types of immunoassay have been applied to the measurement of airborne animal allergens, but all depend on the ability of antibodies to specifically bind relevant animal proteins. Polyclonal assays include those utilizing specific human IgE (such as RAST inhibition, first described by Agarwal *et al.* 1981) and specific IgG antisera raised in rabbits (or other animal species). Assays based on polyclonal antiserum have advantages in that they may measure many relevant airborne proteins, but problems in assay standardization may occur when the antiserum pool expires. The latter problem has been largely overcome with the development of monoclonal antibodies. Recently, assays have been developed using monoclonal antibodies raised to allergens with two distinct, non-overlapping epitopes (Platts-Mills *et al.* 1989). Such two-site assays have been used for the measurement of the major allergens of the house dust mite (Der p 1 and Der f 1) (Chapman *et al.* 1987, Luczynska *et al.* 1989) and the major cat allergen (Fel d 1) (Chapman *et al.* 1988, Lombardero *et al.* 1988). These methods are usually more sensitive than polyclonal assays and have enabled worldwide standardization of mite and cat allergen measurement. However, due to the nature of the monospecific antibody, such assays can measure only one dust constituent, which should therefore be a major allergen which can be used as an appropriate marker. Whilst a suitable marker protein (Rat n 1) has been identified in airborne dust from a rat room (Gordon *et al.* 1996) and a monoclonal antibody assay developed (Renström *et al.* 1996b), this allergen is only one of at least three major allergens and the suitability of quantitation of rat exposure by this method is being investigated (Renström *et al.* 1996a). The identification of marker proteins and the development of assays for other animal species is less well advanced.

The measurement of aeroallergen in the environment is of increasing importance as interest grows in studying the influence of aeroallergen exposure on the development of allergic disease and in controlling occupational allergen exposure. Several methods to measure airborne allergen from rat (Corn *et al.* 1988, Davies *et al.* 1983b, Edwards *et al.* 1983, Eggleston *et al.* 1989, Gordon *et al.* 1992a, Hollander *et al.* 1996, Lewis *et al.* 1988, Platts-Mills *et al.* 1986, Price & Longbottom 1988b, Sakaguchi *et al.* 1989 and 1990, Swanson *et al.* 1985) and mouse (Gordon *et al.* 1997a, Hollander *et al.* 1996, Ohman *et al.* 1994, Twiggs *et al.* 1982) have been described. The comparison of absolute measurements of airborne allergen between animals and between studies is difficult; reported levels for rat allergen, for example, vary from 10^{-10} g/m^3 to 10^{-6} g/m^3. A recent examination of filters and elution methods commonly employed in the measurement of rat urine aeroallergen concluded that the most important factor influencing the results was the inclusion of Tween 20 in the elution buffer. Concentrations of 0.5% Tween 20 enhanced allergen elution up to 10-fold, and this may account for some of the differences in the values reported (Gordon *et al.* 1992b). Other factors that may also contribute to the difference in values reported are the different assay and standards employed (Hollander *et al.* 1999, Renström *et al.* 1999). Assay standardization should now be sought at the earliest opportunity.

Exposure data

Although the units of allergen reported may vary, a number of the general observations about animal allergens are very consistent.

Aeroallergen levels are highest in animal rooms, especially during disturbances such as cleaning out (Edwards *et al.* 1983, Swanson *et al.* 1984, Twiggs *et al.* 1982). Personal exposure when working with animals can be 3–10-fold higher than that of static or background concentration of the room (Davies *et al.* 1983b, Eggleston *et al.* 1989, Gordon *et al.* 1992b, Price & Longbottom 1988b). The rat aeroallergen concentration in undisturbed rat rooms has been reported between 300 ng/m^3 (Edwards *et al.* 1983) to 2.3 ng/m^3 (Eggleston *et al.* 1989), whilst personal samples collected during the handling of rats have measured rat aeroallergen concentrations between 3.6 μg/m^3 (Price & Longbottom 1988b) and 15 ng/m^3 (Eggleston *et al.* 1989). The airborne allergen concentrations reported for the mouse show similar variations (Gordon *et al.* 1997a, Hollander *et al.* 1996) and probably reflect the analagous working practices adopted with these species.

It is also a common observation that rat, mouse and guineapig allergens are predominantly carried on large particles, typically 5–15 μm in aerodynamic diameter. The size distribution of the allergenic dust particles is of importance, as the aerodynamic diameter of the dust particles will determine the site of deposition in the airways, which in turn may influence disease symptoms. In general terms, particles greater than 10–15 μm will be deposited predominantly in the nose (extrathoracic fraction), particles less than 2.5 μm will, potentially, enter the alveoli (respirable fraction) and the intermediate particles (2.5–10 μm) will form the tracheo-bronchial fraction. The distribution of particle sizes measured in the air is dependent upon the conditions of the study; recent disturbance of animal litter will encourage the dissemination of large particles which would settle if undisturbed (Platts-Mills *et al.* 1986). Not surprisingly, up to 30% of animal allergens have also been detected on respirable particles (Corn *et al.* 1988, Swanson *et al.* 1990). The airborne dust in animal houses is usually low and the measurement of total dust is not a surrogate for allergen exposure (Nieuwenhuijsen *et al.* 1994).

Exposure-response

The effect of laboratory animal allergen exposure on disease has been studied in a qualitative manner using estimates of exposure such as the frequency of animal contact (days per week) or job description. The results from cross-sectional studies have been variable; recent studies have found an association between symptoms and exposure (Aoyama *et al.* 1992, Kibby *et al.* 1989) although earlier ones did not (Agrup *et al.* 1986, Beeson *et al.* 1983, Cockcroft *et al.* 1981).

The longitudinal studies undertaken by Botham and colleagues showed a sustained decrease in the incidence of LAA when personal respiratory protection measures were introduced (Botham *et al.* 1987, Botham & Teasdale 1987). This would apparently suggest that LAA is largely preventable if allergen exposure can be sufficiently reduced. However, the recent update on this interesting work has indicated that although the reported incidence of symptoms has remained low, the incidence of those who are sensitized to the animals has remained the same (Botham *et al.* 1995).

Few studies have attempted to quantify rat aeroallergen exposure, and the lack of such measurements has probably contributed to the varying association found between exposure and the prevalence of symptoms. An important study examining the relationship of rat urine exposure on LAA prevalence and incidence rates has been published recently (Cullinan *et al.* 1994). The RUA exposure of several workforces has been measured using the strategy of Corn and Esmen (1979). In this approach to exposure assessment the workforce is grouped or zoned according to the similarity of the job they perform (and presumed similarity of exposure). By measuring the exposure of representative workers within each zone, the exposure of the entire workforce can be estimated. Grouping of the aeroallergen exposure categories into low ($0.04\,\mu g/m^3$) medium ($1.07\,\mu g/m^3$) and high ($31.80\,\mu g/m^3$) demonstrated that the prevalence of symptom outcomes showed an increasing trend with increasing exposure, but that the relationship was only significant for skin symptoms and sensitization (positive skin-prick test to rat urine). This observation may be a manifestation of the so-called 'survivor effect'. Skin symptoms are more readily tolerated than respiratory symptoms and hence those with respiratory symptoms may seek to reduce their subsequent exposure.

Although for epidemiological purposes the exposure data were condensed into three categories, the workforce had been divided into nine exposure groups. The rat urinary aeroallergen concentration associated with the six main job description categories in this study is presented in Figure 1 (Gordon *et al.* 1994a). The filters were prepared using the elution conditions recommended by Gordon *et al.* (1992a) and the aeroallergen concentration measured by RAST inhibition using a rat urine standard. Those involved in the day-to-day care of the rats experienced the highest exposure levels, although measurable rat urinary aeroallergen exposure occurred in all the job description categories studied. For comparison, typical measurements made of the background (or static) rat urinary aeroallergen concentration within rooms containing rats was

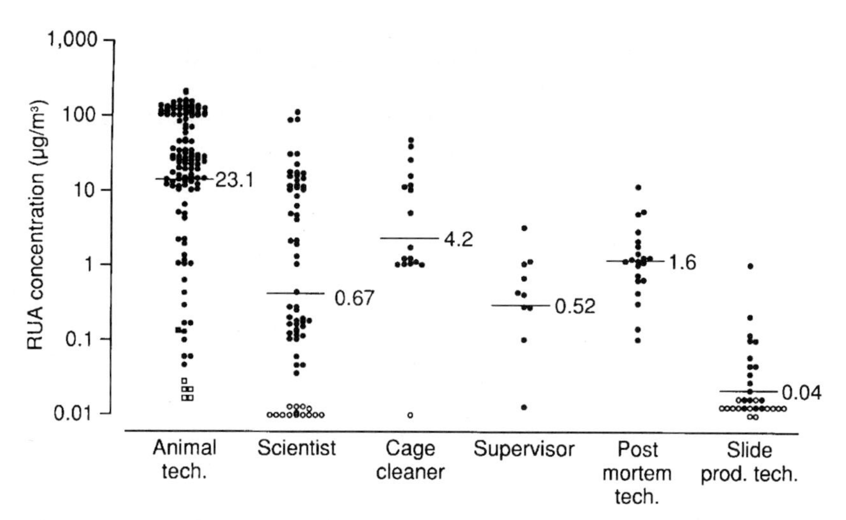

Figure 1. Rat urinary aeroallergen concentration associated with different job descriptions. Circles represent measurements taken with the animals housed in open-top cages, and squares with the animals housed in isolators. Open symbols indicate estimated values at the detection limit of the assay, whilst solid symbols indicate true measurements. The geometric mean is indicated (—). (Reproduced from Gordon *et al.* 1994a, by kind permission)

3.4 µg/m^3 and in the animal house corridors was 0.1 µg/m^3. The variation of RUA concentration within each job description group was large. The factors that contribute to the heterogeneity of the exposure are poorly understood, though the variation in site (ventilation, etc.) and procedures undertaken may account for up to 73% of the variation (Nieuwenhuijsen *et al.* 1995a).

Although the relationship of different exposures with the risk of disease is unknown, the observations of Cullinan and co-workers (1999) suggest that any exposure to rats must be considered to be associated with the risk of sensitization and the development of symptoms. The workers in this cohort have now been followed for five years and the prospective outcomes have been found to be broadly similar to those of the cross-sectional analysis. It was confirmed that symptoms of all types were most common in the first 12 months of exposure and were closely related to a positive rat urine skin-prick test. Eye and nose symptoms were most common, followed by skin symptoms. Interestingly, work-related symptoms tended to be reported before the detection of a positive rat urine skin test. Rat urinary aeroallergen concentration was positively associated with skin, eye and nose symptoms, and to a positive rat urine skin test but not to chest symptoms. These findings were independent of atopic status and of smoking, which were both positively associated with each outcome. Survivor bias was shown to be negligible (Cullinan *et al.* 1999).

Management and prevention of allergy

A general policy statement on safe working practices should be given to all new staff and updated guidance given to existing staff. All staff should be made aware of the allergic hazards and risks of exposure to laboratory animals (Health and Safety Executive 1994), the possible symptoms that can occur and the potentially serious consequences of not reporting symptoms that do develop.

There should be regular monitoring of staff, which should include questionnaires and specific skin-prick tests. These should be carried out on starting employment. Whilst some authors suggest that atopics should be excluded from working with animals (Lincoln *et al.* 1974, Patterson 1964, Schumacher *et al.* 1981), many others claim that too many atopics never develop LAA (Newill *et al.* 1986, Newman Taylor & Gordon 1993). The exclusion of about one-third of the population from employment would therefore not be justified.

Some studies have shown smoking to have an adjuvant effect in the development of LAA, but it is not a discriminating factor. There is no evidence that allergy to other pets predisposes workers to LAA. Neither is there evidence that pre-existing asthma or bronchitis will make workers more vulnerable to LAA. However, if there is airflow limitation due to these diseases and workers develop LAA, they will probably be more severely affected. If allergic symptoms do develop, tests should be carried out monthly to monitor lung function. Continuous medication, e.g. Intal (Rhone-Poulenc Rorer), may be prescribed to control symptoms. If asthma develops the worker should be advised to remove themselves from animal exposure as soon as possible, even if it means the loss of their job.

Although the occurrence of allergy is generally believed to be linked to exposure to the allergen concerned (ABPI 1987, Andersson *et al.* 1990, ESAC 1990), there is to date a lack of documented correlation between building design, e.g. ventilation, and the incidence of disease. However, measures to reduce LAA should be based primarily on environmental control to reduce the concentration of allergen in the air and inhaled by workers and, hence, the prevention of animal and insect excreta and secretions from becoming airborne should be aimed for. Several approaches are required to reduce aeroallergen, including building and ventilation design, animal housing, work routines, personal hygiene and, as a last resort, personal protective equipment (Hunskaar & Fosse 1993). Allied workers, e.g. engineers, drivers, laundry workers, have a potential for being indirectly exposed.

The advent of sensitive methods to measure animal allergen exposure has enabled the effectiveness of intervention studies designed to reduce ambient, or static, allergen concentrations to be monitored. Edwards and co-workers (1983) had demonstrated previously that the level of animal allergens can be reduced by increasing the ventilation rate and relative humidity, but significant reductions were achieved only under conditions that were uncomfortable to work in and detrimental to the health of the animals. Recent improvements in ventilation engineering may, however, have made this a more viable option (Jones *et al.* 1995).

Two studies have demonstrated the importance of soiled litter in the dissemination of rat (Gordon *et al.* 1992a) and mouse allergen (Sakaguchi *et al.* 1990). It is strongly recommended that, where possible, the animals are kept from direct contact with the litter as the use of absorbent pads, not in contact with the animal, is associated with a significant reduction in rat urinary aeroallergen concentrations (Figure 2). Where this is not possible, litter should be selected on the basis of a large particle size (i.e. wood chips rather than sawdust) and probably high absorbency. One of the simplest ways to reduce allergen concentration is to reduce the stock density to a minimum (Figure 3) (Gordon *et al.* 1992b, Nieuwenhuijsen

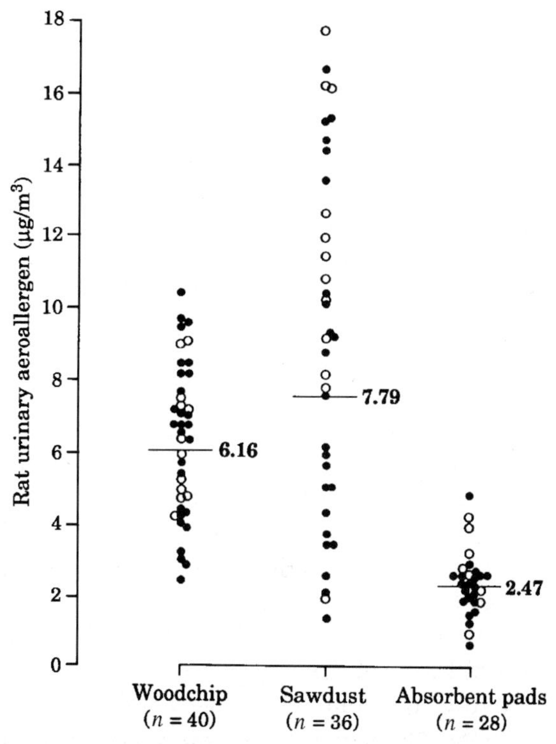

Figure 2. Influence of litter type on rat urinary aeroallergen concentration. Measurements made on cleaning out days are shown as open circles; geometric mean is indicated (—). (Reproduced from Gordon *et al.* 1992b, by kind permission)

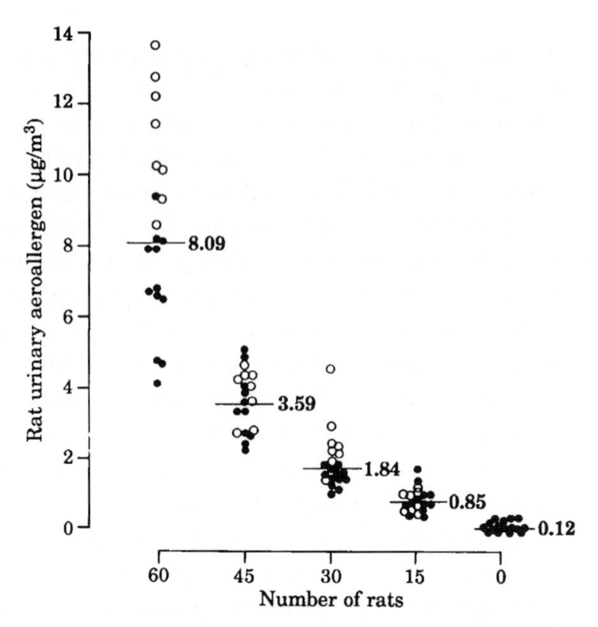

Figure 3. Effect of reducing stock density on rat urinary aeroallergen concentration. Measurements made on cleaning out days are shown as open circles; geometric mean is indicated (—). (Reproduced from Gordon *et al.* 1992b, by kind permission)

et al. 1995b). Where this is not practical, enclosing the animals in flexible film isolators or filter-top cages is an effective way of reducing the allergen concentration. However, because of the potential health consequences to the animals, the use of filter-top cages as a long-term control measure should only be contemplated if extra ventilation of the cages can be provided (Keller *et al.* 1989). The development of cage systems where each cage is individually ventilated enables large numbers of animals to be housed with minimum leakage of allergen into the environment, even if the cages are operated under barrier conditions, i.e. positive pressure to the environment (Gordon *et al.* 1997b).

Recirculation of room air only after filtration through HEPA filters may also have some benefit, providing the generation rate of allergen is not too great (Swanson *et al.* 1990). The use of ventilated work stations for the containment of allergen generated by the handling of animals has been shown to result in a 25-fold reduction of allergen exposure when compared to handling of animals in the open; however, inappropriate use of the ventilated cabinet (such as leaving the front up) reduced the protection factor (Gordon *et al.* 1997b). The same study examined the use of a vacuum system for the removal of soiled litter and found that whilst a reduction in the intensity of exposure was shown, the time taken to complete the task was longer. The safety benefits of such equipment are therefore less clear and the use of additional respiratory protection should be considered.

There have been several reports indicating that personal protective equipment is effective in reducing LAA symptoms. Air-stream respirator helmets have been shown to produce a 75% improvement in asthmatic symptoms (Slovak *et al.* 1985) and in our experience have allowed research workers to continue their studies in the short term. However, it is not easy to do close work wearing these helmets and some people cannot tolerate them. Great care needs to be taken when changing the filters since they represent a major source of allergen; ideally filters should be changed at a ventilated work station. Fine dust masks as used in spray-painting, which cover the nose and mouth, produce a much more efficient fit than simple surgical masks, but again some cannot wear them over long periods as exhaled moisture tends to accumulate on the inside of the mask. The design of face masks has, however, greatly improved in recent years, although no formal studies of their effectiveness at allergy prevention have been published.

Most modern animal facilities require personnel to use specific clothing while working in order to prevent microbiological contamination of both animals and staff. This can range normally from full surgical clothing with gown, cap, mask and shoe covers to a simple laboratory coat change. Care should be taken in coat design not to trap allergens inside the sleeves and thereby increase the risk of contact urticaria on the arms. Hot water soluble laundry bags should be used if contaminated linen has to go out of the unit. Care needs also to be taken to provide separate facilities when people are not working with animals. These should include adequate washing and showering facilities within the animal facility, changing rooms with lockers to isolate non-working clothes and separate office and refreshment rooms.

Work routines in the animal house should be planned in a way that prevents the spread of allergen around the worker. There have been several guideline reports (ABPI 1987, Andersson *et al.* 1990, Bland *et al.* 1987, ESAC 1990). Soiled bedding is the most significant source of allergen in the animal facility and cages should not be emptied in the open room. Dedicated equipment, such as ventilated work stations, should be provided – preferably under negative pressure to minimize contact with the soiled bedding.

The medical consequences of rhinoconjunctivitis or urticaria can be very troublesome, and those with asthma who continue to be exposed can develop increasingly severe symptoms. The severity of symptoms is variable, however, and is dependent on an individual's exposure and tolerance (Eggleston *et al.* 1995). The early detection of LAA will enable precautionary measures to be adopted, such as relocation of the workers or reduction of their exposure, and may prevent progression to severe symptoms (ABPI 1987). Symptomatic subjects should be advised to avoid further exposure to the animals that cause their symptoms. If this is not immediately possible – due, for instance, to completing a research project or not being able to find new employment – continuation in the job may be possible for a short, defined time. However, exposure should be reduced to a minimum, medication may be given and respiratory protection worn. The social consequences of LAA are therefore many and can affect the quality of life (Hunskaar & Fosse 1990). Although immunotherapy is popular in North America for the treatment of hypersensitivity to common environmental allergens, such treatment can carry risks and hence in Britain the pharmacological management of allergic symptoms is currently preferred.

Concluding statement

Exposure to laboratory animals causes approximately a third of those exposed to develop allergic symptoms and about 10% of the workforce to develop asthma. Early diagnosis should be confirmed and where possible the patient removed from exposure.

The reduction of the concentration of airborne allergen inhaled by workers should be aimed for in order to reduce disease incidence. More longitudinal studies are required to give the true incidence rate of disease. With the development of methods to measure aeroallergen, exposure-response relationships can be determined and the effectiveness of measures to reduce aeroallergen can be evaluated. It is hoped that this will lead to standards being recommended for control measures, but it is possible that the use of negative pressure isolators may be the only way to protect totally against the development of LAA.

References

ABPI (The Association of the British Pharmaceutical Industry) (1987) *Advisory note on allergy to laboratory animals* (Dewdney JM, Johnson RD, Skidmore IF, Slovak AJM, Teasdale EL, Williams GA, eds). London: ABPI

Agarwal MK, Yunginger JW, Swanson MC, Reed CE (1981) An immunochemical method to measure atmospheric allergens. *Journal of Allergy and Clinical Immunology* **68**, 194–200

Agrup G, Belin L, Sjöstedt L, Skerfving S (1986) Allergy to laboratory animals in laboratory technicians and animal keepers. *British Journal of Industrial Medicine* **43**, 192–8

Agrup G, Sjöstedt L (1985) Contact urticaria in laboratory technicians working with animals. *Acta Dermato-Venereologica* **65**, 111–15

Alt JM, Hackbarth H, Deerberg F, Stolte H (1980) Proteinuria in rats in relation to age-dependent renal changes. *Laboratory Animals* **14**, 95–101

Andersson S, Baneryd K, Lindh G (1990) *Working with laboratory animals.* Arbetesskydsstyrelsen, Publikasjonsservice, AFS; 11, 29

Aoyama K, Ueda A, Manda F, Matsushita T, Ueda T, Yamauchi C (1992) Allergy to laboratory animals: an epidemiological study. *British Journal of Industrial Medicine* **49**, 41–7

Bacchini A, Gaetani E, Cavaggioni A (1992) Pheromone binding properties of the mouse, *Mus musculus*. *Experientia* **48**, 419–21

Bauer M, Patnode R (1984) Health hazard evaluation report HETA-81-121-1421. West Virginia: NIOSH (National Institute for Occupational Safety and Health)

Baur X (1982) Chironomid hemoglobin. A major allergen for humans. *Chironomus* **2**, 24–5

Bayard C, Holmquist L, Vesterberg O (1996) Purification and identification of allergenic α_{2u}-globulin species of rat urine. *Biochimica et Biophysica Acta* **1290**, 129–34

Beeson MF, Dewdney JM, Edwards RG, Lee D, Orr RG (1983) Prevalence and diagnosis of laboratory animal allergy. *Clinical Allergy* **13**, 433–42

Bellas TE (1990) Occupational inhalant allergy to arthropods. *Clinical Reviews of Allergy* **8**, 15–29

Bland SM, Evans RE, Rivera JC (1987) Allergy to laboratory animals in health care personnel. *Occupational Medicine* **2**, 525–46

Bond HE (1962) A sex-associated protein in liver tissue of the rat and its response to endocrine manipulation. *Nature* **198**, 242–4

Borghoff SJ, Short BG, Swenberg JA (1990) Biochemical mechanisms and pathobiology of α_{2u}-globulin nephropathy (Review). *Annual Reviews of Pharmacology and Toxicology* **30**, 349–67

Botham PA, Davies GE, Teasdale EL (1987) Allergy to laboratory animals: a prospective study of the incidence and of the influence of atopy on its development. *British Journal of Industrial Medicine* **44**, 627–32

Botham PA, Lamb CT, Teasdale EL, Bonner SM, Tomenson JA (1995) Allergy to laboratory animals: a follow up study of its incidence and of the influence of atopy and pre-existing sensitisation on its development. *Occupational and Environmental Medicine* **52**, 129–33

Botham PA, Teasdale EL (1987) Allergy to laboratory animals. *Biologist* **34**, 162–3

Bryant DH, Boscato LM, Mboloi PN, Stuart MC (1995) Allergy to laboratory animals among animal handlers. *The Medical Journal of Australia* **163**, 415–18

Burek JD, Duprat P, Owen R, Peter CP, Van Zwieten MJ (1988) Spontaneous renal disease in laboratory animals (Review). *International Review of Experimental Pathology* **30**, 231–319

Burge PS, Harries MG, Lam WK, O'Brien IM, Patchett PA (1985) Occupational asthma due to formaldehyde. *Thorax* **40**, 255–60

Chapman MD (1993) Dissecting cockroach allergens. *Clinical and Experimental Allergy* **23**, 459–61

Chapman MD, Aalberse RC, Brown MJ, Platts-Mills TAE (1988) Monoclonal antibodies to the major feline allergen *Fel d*1. *Journal of Immunology* **140**, 812–18

Chapman MD, Heymann PW, Wilkins SR, Brown MJ, Platts-Mills TAE (1987) Monoclonal immunoassays for major dust mite (*Dermatophagoides*) allergens, *Der p*1 and *Der f*1, and quantitative analysis of the allergen content of mite and household dust extracts. *Journal of Allergy and Clinical Immunology* **80**, 184–94

Clark AJ, Clissold PM, Shawi RA, Beattie P, Bishop J (1984) Structure of mouse urinary protein genes: different splicing configurations in the 3′ non-coding region. *EMBO Journal* **3**, 1045–52

Cockcroft A, Edwards J, McCarthy P, Andersson N (1981) Allergy in laboratory animal workers. *Lancet* **i**, 827–30

Committee Report (1993) Task force on allergic reactions to latex. *Journal of Allergy and Clinical Immunology* **92**, 16–18

Cormio L, Turjanmaa K, Talja M, Andersson LC, Ruutu M (1993) Toxicity and immediate allergenicity of latex gloves. *Clinical and Experimental Allergy* **23**, 618–23

Corn M, Esmen NA (1979) Workplace exposure zones for classification of employee exposures to physical and chemical agents. *American Industrial Hygiene Association Journal* **40**, 47–57

Corn M, Koegel A, Hall T, Scott A, Newill C, Evans R (1988) Characteristics of airborne particles associated with animal allergy in laboratory workers. *Annals of Occupational Hygiene* **32** (Supplement 1), 435–46

Cullinan P, Cook A, Gordon S, Nieuwenhuijsen MJ, Tee RD, Venables KM, McDonald JC, Newman Taylor AJ (1999) Allergen exposure, atopy and smoking as determinants of allergy to rats in a cohort of laboratory employees. *Eur Respir J* **13**, 1139–43

Cullinan P, Hayes J, Cannon J, Madan I, Heap D, Newman Taylor A J (1992) Occupational asthma in radiographers. *Lancet* **340**, 1477

Cullinan P, Lowson D, Nieuwenhuijsen MJ, Gordon S, Tee RD, Venables KM, McDonald JC, Newman Taylor AJ (1994) Work-related symptoms, sensitisation and estimated exposure in workers not previously exposed to laboratory rats. *Occupational and Environmental Medicine* **51**, 589–92

Davies GE, Thompson AV, Niewola Z, Burrows GE, Teasdale EL, Bird DJ, Phillips DA (1983a) Allergy to laboratory animals: a retrospective and a prospective study. *British Journal of Industrial Medicine* **40**, 442–9

Davies GE, Thompson AV, Rackham M (1983b) Estimation of airborne rat-derived antigens by ELISA. *Journal of Immunoassay* **4**, 113–26

Dinh B-L, Tremblay A, Dufour D (1965) Immunochemical study on rat urinary proteins: their relation to serum and kidney proteins (chromatographic separation of the major urinary protein). *Journal of Immunology* **95**, 574–82

Edwards RG, Beeson MF, Dewdney JM (1983) Laboratory animal allergy: the measurement of airborne urinary allergens and the effects of different environmental conditions. *Laboratory Animals* **17**, 235–9

Eggleston PA, Ansari AA, Adkinson NF, Wood RA (1995) Environmental challenge studies in laboratory animal allergy. Effect of different airborne allergen concentrations. *Journal of Allergy and Clinical Immunology* **151**, 640–6

Eggleston PA, Newill CA, Ansari AA, Pustelnik A, Lou S-R, Evans R III, Marsh DG, Longbottom JL, Corn M (1989) Task-related variation in airborne concentrations of laboratory animal allergens: Studies with *Rat* nl. *Journal of Allergy and Clinical Immunology* **84**, 347–52

ESAC (Education Services Advisory Committee) (1990) *What you should know about allergy to laboratory animals*. Sheffield, UK: HMSO

Finlayson JS, Asofsky R, Potter M, Runner CC (1965) Major urinary protein complex of normal mice: origin. *Science* **149**, 981–2

Finlayson JS, Potter M, Shinnick CS, Smithies O (1974) Components of the major urinary protein complex of inbred mice; Determination of NH_2-terminal sequences and comparison with homologous components from wild mice. *Biochemical Genetics* **11**, 325–35

Fuortes LJ, Weih L, Jones ML, Burmeister LF, Thorne PS, Pollen S, Merchant JA (1996) Epidemiologic assessment of laboratory animal allergy among university employees. *American Journal of Industrial Medicine* **29**, 67–74

Gordon S, Kiernan LA, Nieuwenhuijsen MJ, Cook AD, Tee RD, Newman Taylor AJ (1997a) Measurement of exposure to mouse urinary proteins in an epidemiological study. *Occupational and Environmental Medicine* **54**, 135–40

Gordon S, Tee RD, Lowson D, Newman Taylor AJ (1992a) Comparison and optimization of filter elution methods for the measurement of airborne allergen. *Annals of Occupational Hygiene* **36**, 575–87

Gordon S, Tee RD, Lowson D, Wallace J, Newman Taylor AJ (1992b) Reduction of airborne allergenic urinary proteins from laboratory rats. *British Journal of Industrial Medicine* **49**, 416–22

Gordon S, Tee RD, Newman Taylor AJ (1993a) Analysis of rat urine proteins and allergens by sodium dodecyl sulfate-polyacrylamide gel electrophoresis and immunoblotting. *Journal of Allergy and Clinical Immunology* **92**, 298–305

Gordon S, Tee RD, Newman Taylor AJ (1993b) Comparison of proteins and allergens in rat hair and urine. *Clinical and Experimental Allergy* **23** (Supplement 1), 56 (abstract)

Gordon S, Tee RD, Newman Taylor AJ (1996) Analysis of the allergenic composition of rat dust. *Clinical and Experimental Allergy* **26**, 533–41

Gordon S, Tee RD, Nieuwenhuijsen M, Lowson D, Newman Taylor AJ (1994a) Measurement of airborne rat urinary allergen in an epidemiological study. *Clinical and Experimental Allergy* **24**, 1070–7

Gordon S, Wallace J, Cook AD, Tee RD, Newman Taylor AJ (1997b) Reduction of exposure to laboratory animal allergens in the workplace. *Clinical and Experimental Allergy* **27**, 744–51

Gordon S, Welch JA, Tee RD, Newman Taylor AJ (1994b) Allergenic cross-reactivity between rat and mouse urine. *Clinical and Experimental Allergy* **24**, 176 (abstract)

Grammer LC, Harris KE, Cugell DW, Patterson R (1993) Evaluation of a worker with possible formaldehyde-induced asthma. *Journal of Allergy and Clinical Immunology* **92**, 29–33

Gurka G, Ohman J, Rosenwasser LR (1989) Allergen-specific human T cell clones: Derivation, specificity, and activation requirements. *Journal of Allergy and Clinical Immunology* **83**, 945–54

Hastie ND, Held WA, Toole JJ (1979) Multiple genes coding for the androgen regulated major urinary proteins of the mouse. *Cell* **17**, 449–57

Health and Safety Commission (1999) *General COSHH ACOP and carcinogens ACOP and biological agents ACOP. Control of substances hazardous to health regulations 1999*. L5. Sudbury: HSE Books

Health and Safety Executive (1994) *Preventing asthma at work – how to control respiratory sensitisers*. Sudbury: HSE Books

Heilman DK, Jones RT, Swanson MC, Yunginger JW (1996) A prospective, controlled study showing that rubber gloves are the main contributor to latex aeroallergen levels in the operating room. *Journal of Allergy and Clinical Immunology* **98**, 325–30

Hollander A, Gordon S, Renström A, Thissen J, Doekes G, Larsson P, Malmberg P, Venables KM, Heederik D (1999) Comparison of methods to assess airborne rat and mouse allergens levels I: Analysis of air samples. *Allergy* **54**, 142–9

Hollander A, van Run P, Spithoven J, Heederik D, Doekes G (1996) Exposure of laboratory animal workers to airborne rat and mouse urinary allergens. *Clinical and Experimental Allergy* **26**, 617–26

Home Office (1987) *Statistics of experiments on living animals, Great Britain 1986*. London: HMSO

Hunskaar S, Fosse RT (1990) Allergy to laboratory mice and rats: a review of the pathophysiology, epidemiology and clinical aspects. *Laboratory Animals* **24**, 358–74

Hunskaar S, Fosse RT (1993) Allergy to laboratory mice and rats: a review of its prevention, management and treatment. *Laboratory Animals* **27**, 206–21

Jones RB, Kacergis JB, MacDonald MR, McKnight FT, Turner WA, Ohman JL, Paigen B (1995) The effect of relative humidity on mouse allergen levels in an environmentally controlled mouse room. *American Industrial Hygiene Association Journal* **56**, 398–401

Keller LSF, White WJ, Snider MT, Lang CM (1989) An evaluation of intra-cage ventilation in three animal caging systems. *Laboratory Animal Science* **39**, 237–42

Kibby T, Powell G, Cromer J (1989) Allergy to laboratory animals: A prospective and cross-sectional study. *Journal of Occupational Medicine* **31**, 842–6

Laperche Y, Lynch KR, Dolan KP, Feigelson P (1983) Tissue-specific control of alpha$_{2u}$-globulin gene expression: constitutive synthesis in the submaxillary gland. *Cell* **32**, 453–60

Levy DA, Charpin D, Pecquet C, Leynadier F, Vervloët D (1992) Allergy to latex. *Allergy* **47**, 579–87

Lewis DM, Bledsoe TA, Dement JM (1988) Laboratory animal allergies. Use of the radioallergosorbent test inhibition assay to monitor airborne allergen levels. *Scandinavian Journal of Work and Environmental Health* **14** (Supplement 1), 74–6

Lincoln TA, Bolton NE, Garrett AS (1974) Occupational allergy to animal dander and sera. *Journal of Occupational Medicine* **16**, 465–9

Lombardero M, Carreira J, Duffort O (1988) Monoclonal antibody based radioimmunoassay for the quantitation of the main cat allergen (*Fel d*1 or cat-1). *Journal of Immunological Methods* **108**, 71–6

Lombardi JR, Vandenbergh JG, Whitsett JM (1976) Androgen control of the sexual maturation pheromone in house mouse urine. *Biology of Reproduction* **15**, 179–86

Longbottom JL (1980) Purification and characterisation of allergens from the urines of mice and rats. In: *Advances in Allergology and Immunology* (Oehling A, Glazer E, Arbesman C, eds). Oxford: Pergamon Press: pp 483–90

Longbottom JL (1983) Characterisation of allergens from the urines of experimental animals. In: *Proceedings of XIth International Congress of Allergology and Clinical Immunology* (Kerr TW, Ganderton MA, eds) London and Basingstoke: Macmillan Press: pp 525–9

Longbottom JL, Price JA (1987) Allergy to laboratory animals: characterisation and source of two major mouse allergens Ag1 and Ag3. *International Archives of Allergy and Applied Immunology* **82**, 450–2

Lorusso JR, Moffat S, Ohman JL (1986) Immunologic and biochemical properties of the major mouse urinary allergen (*Mus m1*). *Journal of Allergy and Clinical Immunology* **78**, 928–37

Luczynska CM, Arruda LK, Platts-Mills TAE, Miller JD, Lopez M, Chapman MD (1989) A two-site monoclonal antibody ELISA for the quantification of the major *Dermatophagoides* spp. allergens; *Der p*1 and *Der f*1. *Journal of Immunological Methods* **118**, 227–35

Lutsky I (1980) Occupational asthma in laboratory animal workers. In: *Occupational asthma* (Frazier CA, ed). New York: Van Nostrand Reinhold Co: pp 193–208

Lutsky II, Fink JN, Kidd J, Dahlberg MJE, Yunginger JW (1985) Allergenic properties of rat urine and pelt extracts. *Journal of Allergy and Clinical Immunology* **75**, 279–84

Mancini MA, Majumdar D, Chatterjee B, Roy AK (1989) Alpha$_{2u}$-globulin in modified sebaceous glands with pheromonal functions: localisation of the protein and its mRNA in preputial, meibomian and perianal glands. *Journal of Histochemistry and Cytochemistry* **37**, 149–57

McGivern D, Longbottom JL, Davies D (1985) Allergy to gerbils. *Clinical Allergy* **15**, 163–5

Meredith SK, Taylor V, McDonald JC (1991) Occupational respiratory disease in the United Kingdom: a report to the British Thoracic Society and the Society of Occupational Medicine by the SWORD project group. *British Journal of Industrial Medicine* **48**, 292–8

Morgenstern JP, Griffith IJ, Brauer AW, Rogers BL, Bond JF, Chapman MD, Kuo Mei-chang (1991) Amino acid sequence of *Fel d*1, the major allergen of the domestic cat: protein sequence analysis of cDNA cloning. *Proceedings of the National Academy of Science* **88**, 9690–4

Newill CA, Evand R, Khoury MJ (1986) Preemployment screening for allergy to laboratory animals: Epidemiologic evaluation of its potential usefulness. *Journal of Occupational Medicine* **28**, 1158–64

Newman Taylor AJ, Gordon S (1993) Laboratory animal and insect allergy. In: *Asthma in the workplace* (Bernstein IL, Chan-Yeung M, Malo J-L, Bernstein DI, eds). New York: Marcel Dekker Inc: pp 399–414

Newman Taylor AJ, Longbottom JL, Pepys J (1977) Respiratory allergy to urine proteins of rats and mice. *Lancet* **ii**, 847–9

Newman Taylor AJ, Tee RD (1989) Occupational lung disease. *Current Opinion in Immunology* **1**, 684–9

Nieuwenhuijsen MJ, Gordon S, Harris JM, Tee RD, Venables KM, Newman Taylor AJ (1995a) Variation in rat urinary aeroallergen levels explained by differences in site, task and exposure groups. *Annals of Occupational Hygiene* **39**, 819–25

Nieuwenhuijsen MJ, Gordon S, Harris JM, Tee RD, Venables KM, Newman Taylor AJ (1995b) Determinants of airborne allergen exposure in an animal house. *Occupational Hygiene* **1**, 317–24

Nieuwenhuijsen MJ, Gordon S, Tee RD, Venables KM, McDonald JC, Newman Taylor AJ (1994) Exposure to dust and rat urinary aeroallergens in research establishments. *Occupational and Environmental Medicine* **51**, 593–6

Ohman JL, Hagberg K, MacDonald MR, Jones RR, Paigen BJ, Karcergis JB (1994) Distribution of airborne mouse allergen in a major breeding facility. *Journal of Allergy and Clinical Immunology* **94**, 810–17

Ohman JL, Lowell FC, Bloch K J (1975) Allergens of mammalian origin: II. Characterisation of allergens extracted from rat, mouse, guinea pig and rabbit pelts. *Journal of Allergy and Clinical Immunology* **55**, 16–24

Patterson R (1964) The problem of allergy to laboratory animals. *Laboratory Animal Care* **14**, 466–9

Perlman F (1958) Insects as inhalant allergens. Considerations of aerobiology, biochemistry, preparation of materials and clinical observations. *Journal of Allergy* **29**, 302–28

Petry RW, Voss MJ, Kroutil BS, Crowley W, Bush RK, Busse WW (1985) Monkey dander asthma. *Journal of Allergy and Clinical Immunology* **75**, 268–71

Platts-Mills TAE, Chapman MD, Heymann PW, Luczynska CM (1989) Measurement of airborne allergen using immunoassays (Review). *Immunology and Allergy Clinics of North America* **9**, 269–83

Platts-Mills TAE, Heymann PW, Longbottom JL, Wilkins SR (1986) Airborne allergens associated with asthma: Particle sizes carrying dust mite and rat allergens measured with a cascade impactor. *Journal of Allergy and Clinical Immunology* **77**, 850–7

Price JA, Longbottom JL (1986) Allergy to rabbits I. Specificity and non-specificity of RAST and crossed-radioimmunoelectrophoresis due to the presence of light chains in rabbit allergenic extracts. *Allergy* **41**, 603–12

Price JA, Longbottom JL (1987) Allergy to mice I. Identification of two major mouse allergens (Ag1 and Ag 3) and investigation of their possible origin. *Clinical Allergy* **17**, 43–53

Price JA, Longbottom JL (1988a) Allergy to rabbits II. Identification and characterization of a major rabbit allergen. *Allergy* **43**, 39–48

Price JA, Longbottom JL (1988b) ELISA method for measurement of airborne levels of major laboratory animal allergens. *Clinical Allergy* **18**, 95–107

Price JA, Longbottom JL (1989) Allergy to mice II. Further characterisation of two major mouse allergens (AG 1 and AG 3) and immunohistochemical investigations of their sources. *Clinical and Experimental Allergy* **20**, 71–7

Renström A, Gordon S, Hollander A, Spithoven J, Larsson PH, Venables KM, Heederik D, Malmberg PL (1999) Comparison of methods to assess airborne rat or mouse allergen levels II: Factors influencing allergen detection. *Allergy* **54**, 150–7

Renström A, Gordon S, Larsson PH, Tee RD, Newman Taylor AJ, Malmberg PL (1996a) Comparison of a monoclonal enzyme linked immunosorbent assay (ELISA) and a radioallergosorbent (RAST) inhibition method for aeroallergen measurement. *Clinical and Experimental Allergy* **27**, 1314–21

Renström A, Larsson PH, Malmberg PL (1996b) A new amplified monoclonal rat allergen assay used for evaluation of ventilation improvements in animal rooms. *Journal of Allergy and Clinical Immunology* **100**, 649–55

Reuter AM, Kennes F, Leonard A, Sassen A (1968) Variations of the prealbumin in serum and urine of mice according to strain and sex. *Comparative Biochemistry and Physiology* **25**, 921–8

Ross DJ, Sallie BA, McDonald JC (1995) SWORD '94: surveillance of work-related and occupational respiratory disease in the UK. *Occupational Medicine* **45**, 175–8

Roy AK, Chatterjee B, Demyan WF, Milin BS, Motwani NM, Nath S, Schiop MJ (1983) Hormone and age-dependent regulation of α_{2u}-globulin gene expression (Review). *Recent Progress in Hormonal Research* **39**, 425–61

Roy AK, Neuhaus OW (1966) Identification of rat urinary proteins by zone and immunoelectrophoresis. *Proceedings of the Society of Experimental Biology and Medicine* **121**, 894–9

Roy AK, Neuhaus OW (1967) Androgenic control of a sex-dependent protein in the rat. *Nature* **214**, 618–20

Rudzki E, Rebandel P, Rogozinski T (1981) Contact urticaria from rat tail, guinea pig, streptomycin and vinyl pyridine. *Contact Dermatitis* **7**, 186–8

Sakaguchi M, Inouye I, Miyazawa H, Kamimura H, Kimura M, Yamazaki S (1989) Evaluation of dust respirators for elimination of mouse aeroallergens. *Laboratory Animal Science* **39**, 63–6

Sakaguchi M, Inouye S, Miyazawa H, Kamimura H, Kimura M, Yamazaki S (1990) Evaluation of counter measures for reduction of mouse airborne allergens. *Laboratory Animal Science* **40**, 613–15

Schou C (1993) Defining allergens of mammalian origin (Review). *Clinical and Experimental Allergy* **23**, 7–14

Schumacher MJ (1980) Characterisation of allergens from the urine and pelts of laboratory mice. *Molecular Immunology* **17**, 1087–95

Schumacher MJ, Tait BD, Holmes MC (1981) Allergy to murine antigens in a biological research institute. *Journal of Allergy and Clinical Immunology* **68**, 310–18

Shaw PH, Held WA, Hastie ND (1983) The gene family for major urinary proteins: Expression in several secretory tissues of the mouse. *Cell* **32**, 755–61

Simon FA (1941) Device for rapid performance of skin test by scratch methods. *Journal of Allergy* **12**, 191–2

Sirganian RP, Sandberg AL (1979) Characterisation of mouse allergens. *Journal of Allergy and Clinical Immunology* **63**, 435–42

Slovak AJM, Hill RN (1981) Laboratory animal allergy: a clinical survey of an exposed population. *British Journal of Industrial Medicine* **38**, 38–41

Slovak AJM, Orr RG, Teasdale EL (1985) Efficacy of the helmet respirator in occupational asthma due to laboratory animal allergy (LAA). *American Industrial Hygiene Association Journal* **46**, 411–15

Soparkar GR, Patel PC, Cockcroft DW (1993) Inhalant atopic sensitivity to grasshoppers in research laboratories. *Journal of Allergy and Clinical Immunology* **92**, 61–5

Spieksma FTM, Vooren PH, Kramps JA, Dijkman JH (1986) Respiratory allergy to laboratory fruit flies (*Drosophila melanogaster*). *Journal of Allergy and Clinical Immunology* **77**, 108–13

Stewart GA, Thompson PJ (1996) The biochemistry of common aeroallergens. *Clinical and Experimental Allergy* **26**, 1020–44

Swanson MC, Agarwal MK, Reed C E (1985) An immunochemical approach to indoor aeroallergen quantitation with a new volumetric air sampler: Studies with mite, roach, cat, mouse and guinea pig antigens. *Journal of Allergy and Clinical Immunology* **76**, 724–9

Swanson MC, Agarwal MK, Yunginger JW, Reed CE (1984) Guinea pig derived allergens. Clinicoimmunologic studies, characterisation, airborne quantitation and size distribution. *American Review of Respiratory Disease* **129**, 844–9

Swanson MC, Bubak ME, Hunt LW, Yunginger JW, Warner MA, Reed CE (1994) Quantification of occupational latex aeroallergens in a medical centre. *Journal of Allergy and Clinical Immunology* **94**, 445–51

Swanson MC, Campbell AR, O'Hollaren MT, Reed CE (1990) Role of ventilation, air filtration, and allergen production rate in determining concentrations of rat allergens in the air of animal quarters. *American Review of Respiratory Disease* **141**,1578–81

Teasdale EL, Davies GE, Slovak A (1993) Anaphylaxis after bites by rodents. *British Medical Journal* **286**, 1480

Tee RD, Gordon DJ, Gordon S, Crook B, Nunn AJ, Musk AW, Venables KM, Newman Taylor AJ (1992) Immune response to flour and dust mites in a United Kingdom bakery. *British Journal of Industrial Medicine* **49**, 581–7

Tee RD, Gordon DJ, Hawkins ER, Nunn AJ, Lacey J, Venables KM, Cooter RJ, McCaffery AR, Newman Taylor AJ (1988) Occupational allergy to locusts: an investigation of the sources of the allergen. *Journal of Allergy and Clinical Immunology* **81**, 517–25

Tee RD, Gordon DJ, Lacey J, Nunn AJ, Brown M, Newman Taylor AJ (1985) Occupational allergy to the common house fly (*Musca domestica*): use of immunologic response to identify atmospheric allergen. *Journal of Allergy and Clinical Immunology* **76**, 826–31

Twiggs JT, Agarwal MK, Dahlberg MJE, Yunginger JW (1982) Immunochemical measurement of airborne mouse allergens in a laboratory animal facility. *Journal of Allergy and Clinical Immunology* **69**, 522–6

van Hage-Hamsten M (1992) Allergens of storage mites. *Clinical and Experimental Allergy* **22**, 429–31

Vandoren G, Mertens B, Heyns W, Van Baelen H, Rombauts W, Verhoeven G (1983) Different forms of α_{2u}-globulin in male and female rat urine. *European Journal of Biochemistry* **134**, 175–81

Venables KM, Tee RD, Hawkins ER, Gordon DJ, Wale CJ, Farrer NM, Lam TH, Baxter PJ, Newman Taylor AJ (1988) Laboratory animal allergy in a pharmaceutical company. *British Journal of Industrial Medicine* **45**, 660–6

Walls AF, Longbottom JL (1983a.) Quantitative immunoelectrophoretic analysis of rat allergen extracts. I Antigenic characterisation of fur, urine, saliva and other rat-derived materials. *Allergy* **38**, 419–31

Walls AF, Longbottom JL (1983b) Quantitative immunoelectrophoretic analysis of rat allergen extracts. II Fur, urine and saliva studied by crossed radio-immunoelectrophoresis. *Allergy* **38**, 501–12

Walls AF, Newman Taylor AJ, Longbottom JL (1985a) Allergy to guinea pigs: I. Allergenic activities of extracts derived from the pelt, saliva, urine and other sources. *Clinical Allergy* **15**, 241–51

Walls AF, Newman Taylor AJ, Longbottom JL (1985b) Allergy to guinea pigs: II. Identification of specific allergens in guinea pig dust by crossed radio-immunoelectrophoresis and the investigation of the possible origin. *Clinical Allergy* **15**, 535–46

Warner JA, Longbottom JL (1991) Allergy to rabbits. III. Further identification and characterization of rabbit allergens. *Allergy* **46**, 481–91

Watt AD, McSharry CP (1996) Laboratory animal allergy: anaphylaxis from a needle injury. *Occupational and Environmental Medicine* **53**, 573–4

Weaver RN, Gray JE, Schultz JR (1975) Urinary proteins in Sprague–Dawley rats with chronic progressive nephrosis. *Laboratory Animal Science* **25**, 705–10

Wirtz RA (1980) Occupational allergies to arthropods. *Bulletin of the Entomological Society of America* **26**, 356–60

3

Infectious hazards

M J Dennis

Contents

Introduction

Laboratory animal science is, like other branches of science, constantly progressing. In some ways, however, it has made striking advances over the last decade in terms of increased availability and demand for animals of defined microbiological quality. Most commercial breeders now offer a range of virus-free rodents and most other species can be obtained from barrier-reared, pathogen-free colonies. For various reasons including important ethical ones, there has been a definite swing away from the use of wild-caught animals towards captive-bred. This is particularly the case with primates where various governmental authorities have made the import or export of wild-caught primates much more difficult. There is recognition within the scientific community that a healthy captive-bred animal is a more valuable resource scientifically than one that has suffered the stresses of capture and transportation; this also helps to address the ethical criticism that this trade has invoked.

All these factors, whilst increasing the scientific value of laboratory animals, have also led to a reduction in the hazard that these animals pose in terms of potential infections carried by them which can be naturally transmitted to man (zoonoses). Conversely, scientific progress has also led to developments in animal models which make them more of a hazard to those who use them and are responsible for their care. For example, the technology of transgenics and the development of immunodeficient animals have led to animal models that are more susceptible to a wider range of human pathogens than before (see below). Transplantation of animal organs or tissues to humans may be a source of infection.

Hazards and risks

The general laws and guidelines relating to control of infection, including risk assessment and waste management, are dealt with in some detail in Chapters 7 and 8, and a guide to preparing a microbiological risk assessment is given at the end of this chapter. Infectious hazards and risks arising from the use of laboratory animals must be assessed and appropriate precautions instituted. The following considerations are important in making such assessments.

Laboratory animals, unlike many other laboratory systems, pose a risk not only from the microorganism under study but also from any potentially zoonotic disease agents that they may be carrying. Thus all areas containing animals should be considered as posing some degree of risk even when being used simply to house or breed animals not involved in infectious disease studies. The risks to animal facility users posed by infectious agents can, therefore, be divided into two distinct areas: experimental infections and zoonotic infections.

Experimental infections

The study of infectious agents in animals could be considered to be more hazardous than other laboratory procedures involving micro-organisms. This is because the use of animal models presents many unpredictable variables including the amount of growth and excretion of the organism and the unpredictable behaviour of the animals themselves during the infection process and subsequent sampling procedures.

Categorization of infectious agents

In many countries, including the United Kingdom, systems have been devised to classify infectious agents according to the hazards they present to laboratory workers and the community. Assessment of risks posed by infectious agents is dependent on a variety of factors such as severity of disease, infectivity of the agent, the route by which it infects and the availability of effective prophylaxis or treatment.

In the UK the Advisory Committee on Dangerous Pathogens (ACDP) has categorized biological agents into four hazard groups using the following criteria:

Group 1. A biological agent unlikely to cause human disease.

Group 2. A biological agent that can cause human disease and may be a hazard to employees; it is unlikely to spread to the community and effective prophylaxis or effective treatment is usually available.

Group 3. A biological agent that can cause severe human disease and presents a serious hazard to employees; it may present a risk of spreading to the community but there is usually effective prophylaxis or treatment available.

Group 4. A biological agent that causes severe human disease and is a serious hazard to employees; it is likely to spread to the community and there is usually no effective prophylaxis or treatment available.

Currently a fourth edition (1995) of ACDP's *Categorisation of biological agents according to hazard and categories of containment* lists agents in their various categories, and this approved list implements European Community Directive 93/88/EEC which contains a Community classification of biological agents.

The above criteria for the classification of organisms into biohazard groups are broadly similar in most countries where lists have been drawn up, although the placement of individual organisms may vary. As previously mentioned, the UK and other EC countries are covered by a Community classification of biological agents (EC Directive 93/88/EEC). In the United States this task has been performed by the US Department of Health and Human Services (Biosafety in Microbiological and Biomedical Laboratories, CDC-NIH, 1993), in Canada by the Laboratory Center for Disease Control (Health and Welfare Canada 1990) and in Australia by the Standards Association of Australia (Australian Standard AS 2243, part 3, 1991).

In the UK the Ministry of Agriculture Fisheries and Food (MAFF) Specified Animal Pathogens Order 1998 prohibits any person from having in their possession or deliberately introducing into animals any specified animal pathogen unless they hold a suitable licence to do so. As the primary concern of MAFF is to prevent the spread of exotic pathogens, they may impose higher containment conditions on the use of a particular organism than are required by ACDP recommendations.

When making an assessment of the hazard and risk of infection presented by a particular microorganism, one should bear in mind several factors. The infectivity of the organism for the animal (this may depend on age, species, strain and other factors), its tropism for particular tissues, the likelihood of it being excreted and by what route, the possible route of infection for man and its infective dose for man where this is known are all important considerations. Table 1 indicates the widely differing figures for minimum infectious doses of various organisms in man based on figures provided by the US National Institutes of Health.

The route of infection can determine the infective dose of some microorganisms and under experimental conditions this may not be the natural route for man. For example, the natural route of infection for a particular bacterium may be via small lesions in the skin and may require only a few organisms to cause disease. Under experimental conditions, however, sufficient numbers of organisms may be generated in the environment to allow infection to occur by another route such as the respiratory tract. In many cases the number of organisms required to cause infection via a particular route may vary widely according to species and even to strains

Table 1 Infectious doses of some infectious agents in man

Disease agent	Type	Route of infection	Dose+
Venezuelan equine encephalitis	Virus	Subcutaneous	1*
Parainfluenza 1	Virus	Intranasal	1
Poliovirus	Virus	Ingestion	2
Rubella	Virus	Pharyngeal	$\geq$10
Coxsackie A21	Virus	Inhalation	$\geq$18
Rubella	Virus	Subcutaneous	30
Rubella	Virus	Intranasal	60
SV40	Virus	Nasopharyngeal	10^3
Coxiella burnettii	Rickettsia	Inhalation	10
Rickettsia tsutsugamushi	Rickettsia	Intradermal	3
Francisella tularensis	Bacterium	Inhalation	10
Shigella flexneri	Bacterium	Ingestion	180
Bacillus anthracis	Bacterium	Inhalation	$\geq$1300
Vibrio cholerae	Bacterium	Ingestion	10^8
Escherichia coli	Bacterium	Ingestion	10^8
VTEC 0157	Bacterium	Ingestion	40

+ Dose = No. of organisms required to cause disease in human volunteers or *animal infection extrapolated to man

within a species. In general, it can be seen from Table 1 that viruses and rickettsiae tend to be more infective than bacteria, but *Francisella tularensis* with an infective dose of 10 inhaled organisms shows that there are exceptions.

The use of the concept of virulent and avirulent strains when referring to the ability to invade the body and cause disease should be treated with caution, as these are relative and not absolute terms. Many so-called avirulent organisms can produce severe disease if the dose is sufficiently high and some may be avirulent only in certain species; for example, the SI9 vaccine strain of *Brucella abortus* (Stableforth 1953) protects cattle from disease but may actually cause disease in man. There is also the possibility that an avirulent organism may revert to virulence by passage through an animal host or that virulence factors may be selected preferentially by growth in a particular animal host. The possibility of accidental substitution of virulent for avirulent strains must also be considered.

As mentioned above, recent technologies such as the development of immunodeficient animals, e.g. nude mice and rats, severe combined immunodeficient (SCID) mice, a whole range of germ-free or gnotobiotic species and transgenic animals, has led to animal models which are permissive or susceptible to a wide range of human pathogens. The SCID mouse, in particular, can be engrafted either surgically or by injection with a variety of human tissues which can then be targeted by human infectious agents which normally might not be infectious to mice.

The potential hazard of an infectious agent may also depend on the health and immune status of those entering the animal facility. Vaccinia virus can produce severe disease in persons suffering from eczema even when this condition is quiescent. The normal immunological defence mechanisms may be compromised by congenital or acquired immunodeficiency states, e.g. HIV infection, certain types of cancer or treatment with immunosuppressant drugs or radiotherapy or during pregnancy. Under these conditions even normally harmless organisms may become opportunistic pathogens and give rise to serious disease. In addition, antibiotic therapy may cause an imbalance of the normal microflora which impart some degree of colonization resistance to pathogens in healthy individuals (Grubb *et al.* 1988).

Routes and mechanisms

There are several stages in the process of using animals for infectious disease experiments, all of which present some degree of hazard and risk: the actual infection of the animal, the experimental period during which disease may develop and various samples may be taken and, finally, the autopsy and disposal of the dead animal.

Methods of infection

The most common method of infecting animals is by direct injection using a syringe and needle. This is potentially one of the most hazardous procedures undertaken in the animal facility. Several surveys of accidental infection have incriminated the needle and syringe in around 25% of equipment-related accidents (Sulkin & Pike 1951, Pike *et al.* 1965, Pike 1976). The main ways in which infection can arise are from injection of the operator or their assistant when restraining the animal, from the inhalation of aerosols or from contamination of fingers and the environment.

The risks of accidental injection are obvious and may be due to handling a difficult animal, poor restraint or an inexperienced operator. The use of gloved isolators or safety cabinets and the restrictions that these impose on manoeuvrability could possibly increase the risk of needlestick injury. A review by Collins and Kennedy (1987) identified 50 cases of infection arising from needlestick injury. Many such accidents occur during the practice of disconnecting the needle from the syringe so that, for example, blood may be discharged gently into a container to avoid haemolysis. The risk of needlestick will also occur if needles are unsheathed too soon, are carelessly discarded or attempts are made to replace or resheath them.

Aerosols are produced during several manipulations with syringes and needles. The practice of expressing air bubbles through the needle after filling the syringe is well recognized as being likely to generate aerosols. Not so obvious is the fact that needles will vibrate and produce aerosols when withdrawn through rubber closures such as those on a vaccine bottle. High-speed photography shows that the fine thread of liquid drawn out through the stopper by the tip of the needle breaks up into a string of droplets (Darlow 1972). The smaller droplets become aerosols whilst the larger ones fall to contaminate any surfaces beneath them.

Accidental discharge of the contents of a syringe can occur when the needle is not fitted tightly to the butt of the syringe. Either the needle flies off or, if it is introduced into an animal or a stopper, separates violently from the syringe. Given that the amount of pressure required to raise a bleb in the skin of a guineapig during intradermal inoculation is about 2 atmospheres, it can be seen that such an occurrence will create an extremely dangerous situation. The degree of danger will depend on the infectivity and virulence of the organism being used as well as its concentration. Simulated accidents have also demonstrated that room ventilation plays its part in the dissemination of the aerosol. When 0.5 ml of bacterial culture was discharged into an unventilated room of $50.6\,m^3$ capacity, an average of 235 colonies were collected over a 10 minute air-sampling period. If the ventilation system was switched on at the rate of 11 air changes per hour, an average of only 124 colonies were collected over the same period. There was, however, a much wider dispersal of the organism throughout the room (Hanel & Alg 1955). Thus ventilation systems will generally result in wider dissemination but faster clearance of an aerosol. Aerosols can also be produced by dropping or breaking the syringe and needle, by knocking over an unsupported storage vessel containing the inoculum or even during the process of removing the lid from such a vessel.

Another risk associated with syringes if the plunger is not a tight fit within the barrel is that the contents may leak backwards and contaminate the fingers of the operator via the exposed portion of the barrel. Even with well-fitting plungers there may be sufficient capillary action to cause a leak of infectious material, whilst in some disposable syringes the seal on the plunger may fail after the contents have been drawn up, causing uncontrolled discharge of some of the contents.

The study of respiratory infections in animals poses a particular risk of generating aerosols. The survival of an organism in an aerosol droplet will depend on temperature, relative humidity, the size of the droplet and its chemical content. Experimentally induced variations seem to affect the survival of microorganisms in aerosols. Drying and surface tension effects may be detrimental to survival, whilst other factors such as salt concentration or the presence of proteinaceous material such as sputum or serum may have a protective effect. The size of droplets in aerosols will determine where, if at all, inhaled organisms are deposited in the respiratory tract. Particles 1–3 μm in diameter are predominantly retained in the lower airways, whereas those above 5 μm are usually retained in the nose and pharynx. Using various laboratory animal models it has been shown that if the lower respiratory tract is susceptible to a particular infection, then the number of inhaled particles required to initiate infection rises dramatically with increased particle size. However, if the susceptible region lies in the upper respiratory tract where particles of varying sizes are deposited, particle size effects are less marked. These factors have been reviewed by Dennis (1986).

Intranasal installation can often result in bubble formation on the anterior nares due to exhalation, thus creating an aerosol hazard which will be related to the concentration of organisms in the inoculum. Reitman *et al.* (1954) demonstrated that during the intranasal inoculation of mice with bacteriophage, 27 particles were recovered during inoculation of 10 mice with a 10^8/ml concentration. This was reduced to 0.1 particles per operation using a 10^3/ml dose.

Infection of animals by direct exposure to experimentally produced aerosols can cause hazards even if the actual exposure process is well contained. The aim of this type of experiment is either to reproduce the natural route of infection of a known respiratory pathogen or to confirm a suspected risk from an organism normally considered infective by other routes. In either case, exposure to the aerosol is achieved by either restraining the animal in a device such as the Henderson apparatus (Henderson 1952), in which only the tip of the snout is exposed, or simply exposing the whole animal to the aerosol inside a container (Lillie & Thomson 1972). In the first case the fur around the snout will become contaminated with the organism; in the second, the entire body of the animal will become a potential source of contamination of the environment and of re-infection due to licking its fur and ingesting further amounts of the inoculum. In both cases the animal must be considered to pose a potential hazard both from its body surfaces and from breathing out aerosols of the challenge organism.

Post-infection period

The next stage in an experiment is the holding period after infection, which will certainly involve the animal technician in general day-to-day husbandry and may also involve the technician and scientist in post-infection sampling procedures. After an incubation period, infected animals may begin to excrete the infective agents in the form of exudates, pus or blood from open lesions such as those caused by the use of vaccinia, or from droplets produced by coughing, sneezing, normal respiration or grooming of infected fur. Many infectious agents are also excreted in the urine or faeces and will thus contaminate bedding, fur, cage surfaces and even the walls and floors immediately adjacent to the cage.

Animal technicians should be aware that in carrying out their duties with infected animals they will be confronted by the potential of infectious challenge from aerosols generated by several routes or procedures. Any disturbance of infected bedding can set up infectious aerosols. These will inevitably be caused during cage cleaning or experimental handling of animals, but there will also be peaks of contamination caused by the natural activities of the animals themselves and these peaks will depend on whether the animals are naturally nocturnal, crepuscular or diurnal (Weihe 1975). It has also been demonstrated that dissemination of organisms can occur due to convection currents produced by the body heat of animals (Teelmann & Weihe 1974), but in the context of a modern well ventilated animal room these effects will be minimal.

Infection via the respiratory route can occur from aerosols produced from the animal by exhalation, animal dust, skin scales or fur, dried saliva, urine or faeces. The daily activities of husbandry will create aerosols from disturbed bedding, from sweeping the floor or from infected water droplets created during the washing down of floors or cleaning and re-filling water bottles.

Water droplets or splashes may also cause infection by being swallowed directly or by ingestion via the hands. The hands may become contaminated by direct handling of animals or their waste or even from handling pens, pencils or other articles within the room. Record cards or books may be susceptible to contamination and their movement between 'dirty' and 'clean' areas must be controlled strictly. Other methods of passing information such as networked computer terminals or linked fax machines should be considered.

The skin of the hands and other exposed parts of the body provides an impermeable barrier to most organisms. However, frequently there are small cuts or abrasions sufficiently large to provide entry for microorganisms.

Direct splashes into the eyes also pose the risk of infection via the conjunctival route. Infections reported via this route include *Leptospira* spp. (Welcker 1938), gonococcus (Bruins & Tight 1979), Newcastle disease (Gustafson & Moses 1951), coxsackie V (Dietzman *et al.* 1973), typhus (Wright *et al.* 1968), dengue (Melnick *et al.* 1948) and toxoplasmosis (Field *et al.* 1972).

The direct handling of infected animals is frequently necessary during husbandry procedures such as cage changing or examination of the animals. The added risks from these procedures are of infections via bites or scratch wounds. It should be remembered that animals suffering the symptoms of a disease may be distressed or may become unpredictable or aggressive and should be handled only with extreme caution. Pike (1976) reported that 95 out of 703 infectious accidents involving laboratory workers were caused by bites or scratches.

Experimental sampling of animals will involve several of the risks already mentioned, including bites or scratches, needlestick injury during blood sampling, as well as cuts from scalpels, puncture wounds from suture needles during biopsy or aerosol production during lung lavage, rectal or vaginal washing. The use of gaseous anaesthesia on infected animals may also present a hazard of potential exposure of the operator to an aerosol of microorganisms escaping from leaks in the system, particularly where a mask is used rather than intubation (Schlech 1988).

Necropsy

The final stage in an infectious disease study may involve necropsy and here the risks are from wounds caused by sharp instruments such as scalpels, saws or scissors or from exposed spicules of bone. Aerosol hazards may be caused by splashes of blood, exudates or other body fluids, or when cutting into exposed organs, particularly lung. The scale of the risk here may be estimated from the findings of Pike (1976) that 75 out of 3921 laboratory infections were acquired during autopsies on human cadavers. More recently, Miller *et al.* (1987) found that animal necropsies were the cause of 12 accidental lacerations with instruments, six incidents of splash or spray and four incidents of conjunctival exposure. The handling of potentially infected samples such as blood, organs and other tissues and their subsequent transportation are also hazardous activities.

At the end of any experimental procedure, the disposal of infected carcasses presents a potential risk, particularly if incineration takes place at a site remote from the animal facilities.

Zoonotic infections

As mentioned earlier, the standard of laboratory animals has risen considerably over the last few years and hopefully the days of incidents such as 113 cases of haemorrhagic fever caused by bringing in wild mice and voles for research purposes (Kulagin *et al.* 1962) are long gone. Nevertheless, barrier systems do break down, different species may need to be used as experimental models and species such as farm animals and primates are not available to the same exacting standards as rats and mice. For these reasons workers in animal facilities need to be aware of the potential hazards of zoonotic infection or 'recycled' human infection. All efforts should be made to prevent the entry of wild rodents or ectoparasites, and to isolate all newly introduced animals even if apparently healthy. A report of the National Animal Disease Center (Miller *et al.* 1987) cited 128 cases of exposure to zoonotic organisms between 1960 and 1985, resulting in 34 infections. The highest risk of infection came from *Brucella, Mycobacterium* and *Leptospira* spp.

Infected animals may not necessarily show obvious signs of illness. Viruses, in particular, can persist in their animal hosts for long periods either as a chronic low-grade infection or in a latent non-infectious state which may be reactivated by some form of stress.

A representative sample of zoonotic agents is shown in Table 2 and some of the more important zoonoses will be discussed in relation to individual animal species.

Ironically, the clean status of most laboratory animals can render them much more susceptible to infection from outside agencies.

Primates

Because of the close biological relationship between humans and primates, monkeys are the most dangerous sources of zoonoses amongst the commonly used laboratory species. Some of these zoonotic infections such as *Herpesvirus simiae* (B-virus), filovirus, e.g. the Marburg and Ebola viruses, and rabies virus are known to be particularly hazardous, whilst other simian herpesviruses such as SA8 virus from vervets and the baboon herpes virus should be regarded with caution. A recent report on viral infections of non-human primates as determined by samples submitted for diagnosis indicates the range of viruses to be found in a number of primate species (Kalter *et al.* 1997). Primates may also carry other human pathogens such as mycobacteria, shigellae, salmonellae, poxvirus, protozoa and helminths. Guidance is given in ACDP (1998).

Table 2 Examples of zoonotic infections

Animal source	Organism	Human disease	ACDP category
Primates	Filovirus	Marburg or Ebola haemorrhagic fever	4
	Herpesvirus simiae	Herpes B infection	3
	Rabies virus	Rabies	3
	Hepatitis virus	Hepatitis	3
	Shigella spp.	Dysentery	2
	Salmonella spp.	Gastroenteritis	2
	Mycobacterium tuberculosis	Tuberculosis	3
	Entamoeba histolytica	Amoebic dysentery	2
Rats and mice	LCM virus	Lymphocytic choriomeningitis	3
	Hantaan virus	Korean haemorrhagic fever	3
	Campylobacter spp.	Campylobacteriosis	2
	Streptobacillus moniliformis	Rat bite fever	2
	Spirillum minus	Rat bite fever	1
	Leptospira ballum	Leptospirosis (Weil's disease)	2
	Salmonella spp.	Gastroenteritis	2
	Pseudomonas pseudomallei	Melioidosis	3
	Microsporum and *Trychophyton*	Ringworm	2
Guineapigs	*Salmonella* spp.	Gastroenteritis	2
	Mycobacterium tuberculosis	Tuberculosis	3
	Campylobacter spp.	Campylobacteriosis	2
	Microsporum and *Trychophyton*	Ringworm	2
	Leptospira spp.	Leptospirosis	2
Rabbits	*Salmonella* spp.	Gastroenteritis	2
	Pasteurella spp.	Septicaemia	2
	Microsporum and *Trichophyton*	Ringworm	2
Cats	*Pasteurella* spp., especially *P. multocida*	Septicaemia, meningitis	2
	Bartonella henselae	Cat scratch fever	2
	Toxoplasma gondii	Toxoplasmosis	2
	Toxocara cati	Ascarid infection, possible endophthalmitis due to larva migrans	2
Dogs	Rabies virus	Rabies	3
	Campylobacter spp.	Campylobacteriosis	2
	Leptospira spp.	Leptospirosis	2
	Toxocara canis	Ascarid infection, endophthalmitis	2
	Echinococcus	Hydatid disease	3
	Microsporum and *Trichophyton*	Ringworm	2
Birds	Newcastle disease virus	Conjunctivitis	2
	Chlamydia psittaci	Psittacosis (ornithosis)	3
	Campylobacter spp.	Campylobacteriosis	2
	Salmonella spp.	Gastroenteritis	2
Farm animals	*Coxiella burnettii*	Q fever	3
	Chlamydia spp.	Liver and kidney dysfunction, fetal abnormalities	3
	Orf virus	Dermatitis	2
	Brucella spp.	Brucellosis	3
	Campylobacter spp.	Campylobacteriosis	2
	E. coli 0157	Diarrhoea, kidney failure	3
	Microsporum and *Trichophyton*	Ringworm	2

Careful consideration should therefore be given (notwithstanding ethical considerations) as to whether primates are absolutely essential for a particular experiment.

Herpesvirus simiae

B-virus is an old world simian herpes virus from Asiatic monkeys which had, up to 1987, given rise to 24 reported human infections, most of them (21) fatal (Palmer 1987). More recently, an outbreak in Florida involving three monkey handlers (MMWR 1987) provided evidence of human-to-human infection when the wife of one of the handlers also became infected via a minor skin lesion. Two handlers died and both had suffered bites or wounds whilst handling the same sick monkey. One sick monkey and one healthy monkey were involved and were both found to be excreting B-virus in their saliva. More recently, two monkey handlers who had been bitten became infected, both of them dying (MMWR 1989), and Nanda *et al.* (1990) reported the first demonstration of B-virus infection in the eye of a laboratory technician who had been bitten by a rhesus monkey and subsequently died six weeks after his injury.

Asiatic monkeys with antibodies to this virus probably remain latently infected for life, carrying the viral genome in their trigeminal ganglia (Boulter 1975, Vizoso 1975, Boulter & Grant 1977). By analogy with the human *Herpesvirus hominis*, periodic reactivation of the latent virus is to be expected during pregnancy, severe stress or immunosuppression. In macaques B-virus produces little evidence of disease. There may be vesicles and ulceration of the tongue or oral cavities and occasional conjunctivitis.

The greatest hazards arise from handling rhesus monkeys, particularly from bites or scratches but also from handling their secretions or tissues such as primary kidney cell culture or homogenizing organs. Some human cases of infection have not been associated with physical trauma, so the aerosol route of infection cannot be ruled out. Other species such as cynomolgus macaques and baboons may also carry B-virus.

Whilst the risk of acquiring B-virus is small, stringent precautions should be taken to avoid it. If at all possible only seronegative monkeys should be used for experimental purposes. If seropositive animals or animals of which the antibody status is unclear are kept, then precautions should be taken in both handling and caging to avoid bites or scratches. These animals should not be used for experiments which may result in immunosuppression.

Other herpes viruses

African vervet monkeys may be infected with a virus, SA8, serologically related to B-virus, and one death has resulted from encephalitis following a vervet bite (MRC Simian Virus Committee 1990). Baboons may be infected with a virus closely related to SA8.

Whilst B-virus is not found in New World primates, Melendez (1968) recorded a presumptive human infection with New World herpes virus where a patient who had contact with primates recovered from encephalitis and developed antibodies to *Herpesvirus tamarinus*. Another New World simian herpes virus, *Herpesvirus saimiri*, produces an invasive reticulo-proliferative disease in monkey species other than its natural host, the squirrel monkey; and *H. ateles*, which is carried by the spider monkey, has also been shown to produce malignant lymphoma and leukaemia in other primate species. It is important, therefore, that the possibility of human disease should be borne in mind.

Filovirus infection

The two filoviruses of main concern are Marburg virus and Ebola virus, which are morphologically related but serologically distinct with no cross-protection observed in challenge experiments.

Marburg virus infection in man was first reported in 1967 in Germany and Yugoslavia when it was isolated from cases of haemorrhagic fever in monkey handlers and laboratory workers who had been in contact with vervet monkeys recently imported from Uganda (Martini & Siegert 1971, Gordon-Smith *et al.* 1967). There were 25 primary cases with seven deaths in persons exposed to vervets or their tissues, including material from nephrectomies, autopsies and cell culture. Five secondary cases involved persons in direct contact with a primary case (Simpson 1977). The disease in man presents itself as an acute febrile illness with severe haemorrhage and diarrhoea following an incubation period of 6–9 days. Hennessen (1969) found that the infection spread between monkeys by close physical contact and that the virus was present in blood, urine and saliva of infected animals.

Ebola virus was first described in humans in 1976, when outbreaks occurred in Sudan and Zaire (WHO 1978) identifying over 500 cases. These epidemics were associated with a mortality of over 70% and there was evidence of nosocomial transmission and spread within the community. These were followed by a further outbreak in Sudan in 1979 (Baron *et al.* 1983). More recently, an outbreak of haemorrhagic disease in the Cote d'Ivoire was clearly linked to contact with a wild troop of chimpanzees (Le Guenno *et al.* 1995). The incubation period for Ebola virus is 5–9 days and onset is abrupt with severe malaise, headache, high fever, myalgia, joint pains and sore throat. Death occurs as a result of hypovolaemic shock.

In 1989 Ebola-like virus was detected in cynomolgus monkeys imported into the United States from the Philippines and held in quarantine in Virginia. Several monkeys died and five were found to be positive for Ebola-like virus. This Asian source of Ebola-like filovirus was confirmed when a later shipment of monkeys, received in Pennsylvania and imported from the Philippines, began to show a number of deaths. Ebola-related filovirus was isolated from the liver of one animal. Despite serological evidence of infection in staff handling these animals, no associated clinical illness has been reported (MMWR 1990).

A survey of Asian macaques undertaken in the United States (MMWR 1991) showed that of 9287 specimens, 121 (1.3%) had antibody titres high enough to suggest filovirus infection before importation. Animals that seroconverted were imported from Indonesia, Mauritius, Myanmar and the Philippines. In a serological survey of monkeys from the Philippines, China and Uganda, 43.3% reacted positively with at least one of three different filovirus antigens (Becker *et al.* 1992).

The reassuring findings of Fisher-Hoch *et al.* (1992) are that animals surviving filovirus infection develop a high antibody titre 14–21 days after infection and that this coincides with virus clearance. Healthy monkeys with low titre filovirus antibody may be regarded as uninfected, but a series of tests at intervals would be prudent to check for rising antibody titre.

Human monkeypox

This is a zoonosis occurring in the tropical rain forests of West and Central Africa. It was first discovered in man in 1970 (Breman *et al.* 1980) and up to 1986 a total of 404 human cases had been reported in seven countries (Jesek *et al.* 1988). An animal source was reported in 72% of cases, with human-to-human transmission being found in the remaining 28%. A fatality rate of about 10% was reported. A more recent report has documented further cases of monkeypox in Gabon, with two fatalities (Meyer *et al.* 1991). Nine outbreaks of the virus in captive primates have been recorded since 1958 (Arita *et al.* 1985). A more recent outbreak in Zaire in 1996 (Mukinda *et al.* 1997) has shown more evidence of human-to-human transmission.

Simian retroviruses

Following the discovery of human T-lymphotropic viruses (HTLV-I, HTLV-II) and human immunodeficiency viruses (HIV-1, HIV-2), closely related simian equivalent viruses, STLV (Saxinger *et al.* 1984, Hayami *et al.* 1984) and SIV (Letvin *et al.* 1983, Henrickson *et al.* 1983, Gardner *et al.* 1984, Daniel *et al.* 1985) have been reported. SIV are clearly related to HIV, having a tropism for CD4 + ve T-lymphocytes and producing typical AIDS-like symptoms such as weight loss, decrease in T4 lymphocytes and the development of opportunistic infections in infected Asian macaques.

Serological surveys have shown that SIV are found in wild populations of a variety of African primate species, including African green monkeys, sooty mangabeys, Sykes' monkeys, mandrills and chimpanzees (Hayami *et al.* 1994). These viruses do not normally cause disease in the natural host but most are capable of causing AIDS-like disease when given experimentally to macaques such as rhesus and cynomolgus. No evidence of naturally occurring infection has been found in Asian or New World primates.

Whilst there is no evidence of the development of SIV infection in man, there is one reported case of seroconversion to SIV following a needlestick injury (Khabbaz *et al.* 1992).

Other virus infections

Other virus diseases capable of transmission from monkey to man include rabies, yellow fever, Kyasanur Forest disease and the pox viruses Yaba and tanapox. A simian papovavirus, SV40, has also been shown to be infectious to man, but no adverse symptoms have been reported despite large numbers of human infections occurring due to contamination of poliomyelitis and other vaccines (Fraumeni *et al.* 1970, Mortimer *et al.* 1981). A recent survey of breeding monkeys in Japan found that 89% of monkeys at one institute had antibody to SV40 (Asai *et al.* 1991).

It is important to realize that monkeys can acquire human infection from their handlers and may subsequently pass their diseases back to persons in contact with them.

There is evidence of increased incidence of hepatitis among persons handling chimpanzees and this is probably the result of a human-to-monkey-to-human chain of events. Furthermore, there is experimental evidence of the susceptibility of two marmoset species (*Saguinus mystax* and *S. labiatus*) and cynomolgus macaques to human non-A, non-B hepatitis (Tabor 1989, Gupta *et al.* 1990) and of rhesus macaques and vervet monkeys to hepatitis A (Lapsin & Shevtsova 1990, Zamiatina *et al.* 1990). Collins (1993) cites three incidents of infection with hepatitis B virus associated with work with primates and involving 48 individuals in total.

Other human infections such as measles and *Herpes simplex* can cause fatal disease in both New World and Old World primates (Levy & Mirkovic 1971, Gross 1983).

Besides diseases acquired directly from man, monkeys may also be susceptible to viruses from other animal species. There are several reports of outbreaks of callitrichid hepatitis, an arenavirus which causes a fatal disease in tamarins and marmosets and which shows a close antigenic relationship to lymphocytic choriomeningitis virus (LCMV) (Stiglmair-Herb *et al.* 1992, Stephenson *et al.* 1991). Indeed, an outbreak of hepatitis amongst marmosets and tamarins has been attributed to a single feeding of the primates with newborn mice infected with LCMV (Montali 1993).

Bacteria

One of the most important bacterial hazards involving primates is tuberculosis. Tribe (1969) considered tuberculosis to be present in about 10% of imported rhesus monkeys. A more recent survey showed that 17 out of 249 shipments into the United States had monkeys with tuberculosis infection. Sixteen out of these 17 shipments contained cynomolgus macaques and one monkey handler was reported to have developed a positive tuberculin skin test since working with these animals (MMWR 1993).

Two other bacterial diseases of importance are salmonellosis and shigellosis. *Shigella* species cause significant disease in monkeys and appear to spread more readily from monkey to man than *Salmonella* spp. In a recent incident reported by Kennedy *et al.* (1993), a small cluster of dysenteric illness due to *Shigella flexneri* was found amongst technical assistants at a primate research unit. All of the affected individuals had been in regular contact with a colony of cynomolgus macaques, some of which were excreting the organism. Investigation of working practices revealed the potential for infection via the faeco-oral route. Other bacteria known to be carried by primates and capable of causing disease in man include *Yersinia* spp. and *Campylobacter* spp.

Parasites

A number of intestinal parasites capable of infecting man may be carried by monkeys. Helminths such as strongyloid species may be found in rhesus macaques and *Oesophagostomum apiostomum* is common in macaques, baboons and great apes. Three important intestinal protozoa are *Entamoeba histolytica* which causes amoebic dysentery, *Giardia intestinalis* and the large ciliate *Balantidium coli*. *Cryptosporidium* spp. are also commonly found in primates.

Rats, mice, other rodents and rabbits

The extensive availability of full barrier-reared and virus-free rats, mice and guineapigs from a wide range of accredited breeders, most of whom provide a full health history of their breeding colonies, means that there should be no problems with zoonotic agents in the use of these species. However, it should be borne in mind that allowing animals of high microbiological quality to come into contact with other rodents such as hamsters or gerbils, which have not been bred to such exacting standards, to mix them with rats and mice from 'in-house' breeding colonies that have not undergone regular screening, or to house them in conditions that may allow contact with wild rodents, may result in extensive infection with potentially zoonotic agents.

The increased use of xenografts of tissues from other species, especially man, in immunodeficient mice also presents the hazard of unsuspected human disease being passed on to those in daily contact with such animals. The passaging of, for example, ascitic tumours in mice may also be an unwitting means of transmission of infection.

The most hazardous virus to man found in rodents is lymphocytic choriomeningitis virus (LCMV). Most human infections are mild with influenza-like symptoms, muscular pain and respiratory involvement. In some cases, however, meningitis, meningoencephalitis and central nervous system infection may occur. In a recent outbreak investigated by Mahy *et al.* (1991), an animal caretaker developed viral meningitis having worked with nude mice infected with multiple tumour cell lines which were naturally infected with LCMV. Eight workers were found to have been infected with the virus and the report concluded that these workers were working more frequently with nude mice than were other animal handlers. The number of nude mice used had increased five-fold in this facility in the previous year.

The main reservoir of LCM virus is wild mice but other species such as guineapigs, hamsters, rabbits, dogs and primates may become infected, and infected hamsters are reported to have high levels of excretion (Biggar *et al.* 1977). The virus is shed by carrier animals in their urine and saliva, and infection of humans is probably by inhalation of dust from infected bedding or skin contact with urine.

Laboratory rats can carry Hantaan virus which causes Korean haemorrhagic fever in man (Umenai *et al.* 1979, Desmyter 1983) and it has been shown that this can be acquired from immunocytomas propagated in rats (Lloyd & Jones 1986). The risks associated with the exchange of rats or rat tumours between laboratories has been discussed by McKenna *et al.* (1992). The increasing threat of rodent-associated hantavirus infection has been recently highlighted (Bignall 1995).

The disease known as rat bite fever is caused by two organisms – *Spirillum minus* and *Streptobacillus moniliformis*. The symptoms produced by these two agents are similar, with recurrent fever, joint and muscle pains, lymphadenopathy and rash. The difference is that the incubation period with *Sp. minus* is much longer (several weeks compared with a few days for *S. moniliformis)*. Virtually all cases of human infection in animal facilities are caused by *S. moniliformis*. The main source of infection is implied by its name, but there is at least one reported case in the UK of 'rat bite fever without the bite' (Fordham *et al.* 1992). The organism will cause disease in guineapigs (Kirchner *et al.* 1992) and in certain strains of mouse – the C57BL/6J inbred strain being very susceptible (Wullenweber *et al.* 1990, Kaspareit-Rittinghausen *et al.* 1990).

Mice may become infected with *Salmonella* spp. (Tuffery & Innes 1963), usually *S. typhimurium* or *S. enteritidis*. *Yersinia pseudotuberculosis* can be found in guineapigs and rabbits, and may cause disease in man ranging from mesenteric lymphadenitis to severe septicaemia. Rabbits may also harbour endemic *Pasteurella* spp., particularly *P. multocida*, and infection of man may follow a bite or a scratch. Species of *Leptospira* have been isolated from laboratory mice, rats, guineapigs and hamsters, and cases of human leptospirosis (Weil's disease) acquired from such sources have been recorded.

Rodents and rabbits can carry the fungi causing ringworm, *Trichophyton mentagrophytes* and *Microsporum* spp., although infection is often not apparent.

Intestinal parasites of rodents that can be transmitted to man include the tapeworm *Hymenolepis nana* in mice, guineapigs and hamsters and protozoal parasites such as *Giardia* spp. and *Cryptosporidium* spp.

Cats and dogs

The most important virus disease transmissible by dog to man is rabies. Even in countries where the disease is endemic, care should be taken that imported animals are kept isolated from existing groups of animals. The incubation period in dogs is usually 20–60 days but may be several months. Early signs of the disease are dullness, lethargy and a tendency to seek the darker areas of the pen. This is followed after a few days by restlessness and excitability, passing, 3–4 days later, onto the paralytic stage with drooling saliva, ataxia, convulsions and finally death. In some cases the paralytic stage may develop without the preceding excitability. In cats the disease is similar but follows a shorter course.

Other viruses that may be present are LCM in dogs and the Catscratch fever agent in cats.

Werth (1989) has indicated a risk from dogs and cats of infection with *Chlamydia psittaci* and *Coxiella burnettii*.

Pasteurella multocida appears to be part of the normal buccal flora of domestic cats and dogs and there are many reported cases of infection caused by bites or scratches, particularly by cats. Symptoms range from septic arthritis (Guion & Sculco 1992, Kumar & Kannampuzha 1992, Chevalier *et al.* 1991) to meningitis (Dammeijer & McCombe 1991). In addition, infections with *P. pneumotropica* (Minton 1990), *Capnocytophaga canimorsus* (Tjong & van Zeijl 1991, Weber & Hansen 1991) have been reported following dog bites. With regard to cats, feline pox (Mayr 1993) and bacillary angiomatosis-pelionis putatively caused by *Rochalemaea* species (Tappero *et al.* 1993) have been put forward as new zoonoses. Cats have recently also been identified as carriers of *Helicobacter pylori*, which has been found in saliva and faeces. *H. pylori* is associated with peptic ulcers and stomach cancer in man. Several authors have reported the prevalence in commercially reared dogs of *Campylobacter* spp., which are pathogenic for humans (Fox *et al.* 1988, Blaser *et al.* 1980).

Dogs are a natural reservoir of pathogenic *Leptospira* species, particularly *L. icterohaemorrhagiae* and *L. canicola*. Schmidt *et al.* (1989) have reported that there is a chance of infection from dogs that are asymptomatic and have been vaccinated. Human leptospirosis is characterized by sudden onset of fever, headache, muscular and abdominal pain, nausea and vomiting.

Both cats and dogs may carry fungi capable of infecting man – for example, *Microsporum canis*.

The protozoan *Toxoplasma gondii* occurs in many animal species including dogs and cats; the latter are known to shed oocysts from the gut. Only a small proportion of infected people develop disease, but toxoplasmosis may be severe, even fatal, in immunosuppressed individuals and primary infection during pregnancy can lead to congenital infection with serious consequences. Infection of man can follow the ingestion of oocysts present in the excreta, as may infection with another coccidial parasite, *Cryptosporidium parvum*

(Dubey 1993). Links have been made between dogs and cats and *Giardia* infections in man (Dubey 1993, Arashima *et al.* 1992). Other parasites of cats and dogs potentially transmissible to man include the cestode *Echinococcus granulosus*, producing hydatid disease, the nematodes *Toxocara canis* and *T. cati*, and the mange-producing *Sarcoptes* mite.

Birds

The three main infections that can be contracted from birds are Newcastle disease, psittacosis and salmonellosis.

The incidence of Newcastle disease in birds varies widely, human infection probably occurring from the stirring up of dried infected excrement and dust. In man the disease presents as a conjunctivitis, often associated with headache, malaise and chills.

Psittacosis, or ornithosis, is caused by *Chlamydia psittaci* and is often present as an inapparent infection in older birds, whilst younger birds can show signs of diarrhoea and nasal discharge. These can readily infect man, giving rise to severe headache, generalized muscular pains and respiratory involvement. The disease is most severe when contracted from birds of the parrot family, but turkeys are the main carriers amongst farm birds.

The prevalence of *Salmonella* spp., particularly *S. enteritidis*, has increased in many chicken flocks. Under open or pen housing there is the chance of spread amongst birds by ingestion of faeces. Under more intensive rearing conditions it has been shown that infection can occur via the aerosol route (Baskerville *et al.* 1992). Some infections can lead to bacteraemia and spread of the organism to various organs including the oviduct, where infection of eggs can take place. Some inbred strains have shown pronounced differences in susceptibility to *Salmonella* infection (Blackwell 1988).

Mycobacterium avium has also been cited as a zoonotic pathogen which can be spread to man and other species by the aerosol route (Mayr 1993).

Farm animals

A number of viruses capable of producing human disease are present in farm animals. Sheep may carry *Parapoxvirus ovis*, the cause of contagious pustular dermatitis or orf, which gives rise to painful lesions on the fingers; the infection can also be acquired from goats. Sheep in certain areas may be infected with louping-ill virus, a tick-borne infection which results in a biphasic disease in both sheep and man; a febrile period is followed by involvement of the central nervous system. It is thought that infections of man are acquired through the respiratory route and this presents an obvious hazard if experimental intranasal inoculation of animals is undertaken.

Aujeszky's disease virus or pseudorabies can affect both pigs and cattle, and a recent outbreak in these species has been linked to a pruritis of the palms of workers associated with the farm in question (Anusz *et al.* 1992).

A cautionary note should be added here about the transmissible spongiform encephalopathies. Scrapie has been present in sheep for more than two centuries, and more recently bovine spongiform encephalopathy (BSE) has emerged in cattle in the UK. Whilst there is no direct evidence of transmission of these agents to man under normal conditions of contact, the possible hazards of dealing with naturally or experimentally infected animals, their tissues or products incorporating their extracts (bovine serum, animal diets) should be borne in mind. The ACDP publication (1994) *Precautions for work with human and animal transmissible spongiform encephalopathies* should be consulted for advice on risk assessment and categorization of the agents involved.

The rickettsial disease Q-fever may be acquired from infected cows, sheep and goats. The causative organism, *Coxiella burnettii*, is excreted in body fluids, can survive for long periods in soil and is carried via dust from infected animals. Human infection results from inhalation. Usually the disease is similar to a mild influenza but can be severe, with hepatitis or endocarditis. Pregnant animals are a particular risk as the organism builds up to large numbers in specific tissues such as the uterus, placenta, mammary glands, birth fluids and milk. There is thus a danger from aerosols produced during birth, milking or surgery. Hamadeh *et al.* (1992) reported an outbreak of Q-fever associated with the use of sheep placentas for research purposes. Also of concern is ovine chlamydiosis caused by *Chlamydia psittaci*. The organism causes enzootic abortion in ewes and may cause serious infection in pregnant women who, therefore, should not work with sheep or handle the clothing and other items of people who do (Hadley *et al.* 1992). There is also a report of severe infection resulting in abortion following contact with a herd of goats. The organism can also cause conjunctivitis in man (Tontis & Zwalen 1991).

The main bacteria that may be passed to man from farm animals include *Mycobacterium tuberculosis*, *Bacillus anthracis*, *Brucella*, *Salmonella*, *Campylobacter*, *Streptococcus* and *Staphylococcus* spp. It should be noted, bearing in mind the recent outbreaks of human disease, that verocytotoxin-producing *Escherichia coli* of the serotype 0157 have been isolated from apparently healthy cattle in a number of countries (Advisory Committee on the Microbiological Safety of Food 1995).

In terms of other infections, toxoplasmosis can be acquired from sheep, ringworm can be acquired from cattle, and the liver fluke, *Fasciola hepatica*, of cattle and sheep can sometimes be found to infect the human liver. Flukes are transmitted via a secondary host, a free-living snail, and human infection probably occurs by eating green herbage such as watercress contaminated by such snails.

Other species

Other animal species which may be found in the animal facility and which may carry potential human pathogens include amphibia and reptiles, which may carry species of *Salmonella*; and bats which pose a risk of rabies, particularly with vampire bats obtained from endemic areas. Invertebrates such as flies, ticks and snails are well known as vectors of infectious agents and any wild caught invertebrates should be treated as potential carriers of the agent they are known to transmit.

Countermeasures

Containment

The primary objective in the management of both experimentally infected animals and those that present a risk of zoonotic infection is containment. Containment of infection can be addressed at different levels: the first objective should be primary containment of the infection within the animals and their immediate cage environment. This can be achieved by the use of flexible film or rigid isolators which are kept ventilated by HEPA filtered air, the flow of which is adjusted to provide negative pressure within the isolator at all times. This system is effective in that it provides total isolation of the animals throughout the experimental period as all experimental and husbandry procedures are conducted via gloved sleeves built into the isolator walls. Entry and exit of materials is usually via an entry port, which can be disinfected by chemical agents or can even have more sophisticated locking ports and supply drums which obviate the need for chemical disinfection. The disadvantages of this system are that they are time-consuming to work in and restrict the manoeuvrability of the operator, particularly for experimental procedures that require delicate manipulation. Freedom of movement can be improved by introducing half-suits as an integral part of the isolator wall. These improve the reach of the operators and hence their ability to coordinate their movements for procedures such as restraint and inoculation. Half-suits will require their own ventilation system which will render them at positive pressure to the isolator. Because of their self-containedness, isolators have the advantage that a single room can be used for several different experiments at one time, during which the operator does not have to wear elaborate and restrictive protective clothing.

Other commercial systems available include ventilated environmental cabinets or laminar flow shelving systems, which contain the infectious hazard whilst the cages are inside them. They have the disadvantage that the barrier must be broken to conduct any husbandry or procedures, thus giving an opportunity for contamination of the operator or environment whenever the cages have to be removed. This problem may be addressed to some extent by the use of cages with filter-top lids, which can then be transferred to laminar flow or microbiological safety cabinets for any procedural or husbandry work. These filter-tops, however, were designed mainly to protect disease-free animals and the material used would not pass any validation tests against microorganisms. Filter-topped cages also have the further disadvantage that there tends to be a build up of ammonia, temperature and humidity. This type of system can play an important role in reducing the level of contamination within a room, but their limitations must be borne in mind when drawing up a code of practice, particularly where dangerous pathogens are concerned. It is necessary in such cases to address the requirement for additional protection of personnel, including respiratory protective equipment (RPE), but this should be a final step when all other protective measures are considered to be impracticable or insufficient to prevent exposure.

The same principle of reducing the risk of contamination should be applied to *post mortem* examination, and the use of down-draught tables, laminar flow hoods or totally enclosed safety cabinets should be considered.

The second level at which containment should be addressed is retention of the infection within the room (secondary containment). This is achieved by a number of measures involving both design and practice. Rooms containing quarantined animals or those with experimental infections should be physically isolated from all other areas, with access limited to relevant personnel. Access to infectious areas should be through a changing room where strict rules concerning removal of outer clothing and wearing of protective clothing should apply.

The design of animal rooms should take into consideration the management of all aspects of health and safety. Internal surfaces, including floors, walls, ceilings and work surfaces should be easy to keep clean and to decontaminate. This requires them to be smooth, impervious and resistant to both chemical and physical damage. Any equipment or fittings should lend themselves to easy decontamination by liquid or gaseous disinfectants and the rooms and associated air handling systems should be fully sealable to allow complete isolation during the decontamination process. It is good practice to design rooms so that routine servicing such as the changing of light fittings, adjusting airflows, temperatures and humidities and the initiation of fumigation can be performed without entering the rooms. Filter systems should be designed so that they can be changed easily and safely. This requires that filters are accessible and that potentially contaminated filters can be safely contained upon removal. Thought must also be given to the safe disposal of any liquid waste generated during the maintenance of potentially infected animals. The treatment of this type of waste is considered later, but the design of a suitable drainage system must include consideration of the volume of waste, its particulate and chemical nature, and how any problems such as blockage or leakage can be prevented or dealt with if they occur. The containment of large animals requires special consideration if large volumes of slurry are expected to be generated.

Ideally, rooms should be maintained at negative pressure with regard to the access corridor and any air lock or changing room associated with it. This serves to maintain a gradient of pressure (and thus a direction of air flow) from clean to dirty areas. It may be necessary to have different negative pressures between adjacent rooms where different categories of pathogen are used. Where dangerous human pathogens are present, the room should have input and extract air provided by mechanical ventilation. Air handling systems must have dampers and interlock systems to prevent positive pressurization of the room in the event of failure of the extract fan. Extracted air can be passed through HEPA filters, which ideally should be situated at the point of exit from the room and not in some remote plant room.

A regimen for the safe removal of bedding, solid animal waste, carcasses and other waste for decontamination should be written into a code of practice. Usually waste material is either suitably bagged or placed in some other container. In either case the outer surfaces must be decontaminated before removal to an autoclave validated to deal with the 'worst possible load', bearing in mind that steam must permeate the whole load. Minimum values for 'make safe' runs in an autoclave are as follows (given temperatures are as sensed *in the load itself* and extra time is usually allowed as a margin of safety):

Pressure	*Temperature*	*Sterilization time*
15 lb/in^2 (101 kPa)	121°C	15 minutes
20 lb/in^2 (138 kPa)	126°C	10 minutes
30 lb/in^2 (203 kPa)	134°C	3 minutes

Various methods are available for monitoring each load to check that sterilization temperature has been achieved for an appropriate length of time. These include indicator tubes in which the contents either melt or change colour when the load reaches the correct temperature, and spore strips which have to be cultured after autoclaving to demonstrate that sterilization has taken place. It must be borne in mind, however, that interpreting colour changes is subjective and prone to human error. It is good practice to attach a strip of heat-sensitive indicator tape to each item so that unprocessed containers are not opened in error. Where autoclaves serve an infected area they should ideally be double-ended and form part of the barrier between 'clean' and 'dirty' areas: this way there should be confidence that any load which ends up on the 'clean' side has been made safe. Once made safe by autoclaving, waste material and carcasses are best disposed of by incineration. Untreated infected material can be dealt with directly by incineration, but such materials must be securely sealed in leakproof bags for safe loading, and the incinerator must be provided with after-burners for added safety. Incinerators must now by law comply with national or international guidelines on chamber temperatures and levels of emissions of certain waste gases. The rules that apply will depend on the capacity of the incinerator and the nature of the material to be disposed of. Waste disposal regulations are dealt with in Chapters 7 and 8.

Effective washing of caging, racks, water bottles and other reusable equipment may also play an important role in the management of infectious hazards. The use of a cage washer with a very hot rinse cycle and strong detergents will help to prevent cross-infection between loads, particularly where no other disinfection process has been employed.

Depending on the level of containment required it may also be necessary to treat any liquid effluent before discharge into the sewage system. This is best achieved by collection into holding tanks which, when full, can be heated to a temperature sufficient to kill the particular microorganism expected to be present, or by chemical treatment of each batch. It is helpful to have a reserve tank for use while the main tank is being treated and to have some means of safely collecting samples for sterility checks to validate that it is safe to release the treated material. Where chemical methods are used it may be necessary to neutralize the chemical employed in order to prevent environmental damage when the waste is discharged to the sewage system.

The third level of containment is the prevention of the spread of infection to staff entering the rooms containing the infected animals, and preventing the staff from spreading the infection from one room to the next or elsewhere in the facility or into the wider environment.

The key to the safe use of infectious agents in animals is good management. Staff should be trained in safety techniques as applied to animal handling and husbandry, use of appropriate disinfectants, protective clothing and other equipment. They should also be given a good general background in microbiological awareness so that they have some concept of the different types of microorganisms and their properties. Staff who handle animals on a daily basis should be made aware of the potential symptoms of any infectious agent given to animals in their care. It is a good idea to document all training, whether formal or informal. This way staff build up a comprehensive training record; any gaps in knowledge can be pinpointed and action taken accordingly.

Careful thought should be given to the types of emergencies that may arise, and plans to deal with accidental exposure, fire or physical injury in the presence of infectious agents must be drawn up in a suitable protocol. All staff who work in the facility and any support staff who might be called upon in an emergency (e.g. Occupational Health, Safety Officers and Fire service personnel) should be made aware of and trained to follow emergency procedures.

It is often a wise precaution to vaccinate staff against the relevant microorganisms where this is available, though this should be seen only as an added precaution and not a replacement for good practice. Staff working in quarantine areas should be considered for vaccination against any pathogen that is thought likely to occur, bearing in mind the species and country of origin. In addition, all staff should be screened as to their medical history so that any allergies, immunosuppression or other health problems can be identified before they are exposed to pathogenic agents. Further monitoring of staff at risk may include the assessment of antibody levels after vaccination, testing appropriate samples for the presence of pathogens and regular general health screening.

Vaccination

Any requirement for, and the feasibility of, vaccination should be assessed well before an experiment is due to start and be included in the Code of Practice when it is drawn up. Vaccines available include those against the following organisms:

Bacteria	**Viruses**
Bacillus anthracis	Hepatitis A and B
Bordetella pertussis	Louping-ill
Clostridium botulinum	Measles
Clostridium tetani	Mumps
Corynebacterium diphtheriae	Monkeypox
Mycobacterium tuberculosis	Polio
Salmonella typhi	Rabies
Yersinia pestis	Rift Valley fever
	Rubella
	Tick-borne encephalitis
	Venezuelan equine encephalitis
	Yellow fever

Protective clothing

Protective clothing plays an important role in the management of infectious hazards. The minimum requirement for entry to areas housing infected animals should be for staff to remove all outer clothing and put on a complete set of protective clothing such as a lightweight suit, boots, hat and gloves. A changing room providing the only means of entry for personnel should be appropriately sited for this purpose and access to areas containing potential infectious hazards should be restricted to authorized personnel.

Once inside the facility it may be considered necessary for extra protective clothing to be worn when entering rooms where infected animals are kept. The degree of protection required will depend on the nature of the agent, the species used and the type of procedures carried out. It may be sufficient to put on a protective gown which can be discarded on leaving that particular room, with the added safeguard of a disinfectant footbath to decontaminate footwear. If there is a risk of infectious splashes, such as during procedural work or necropsy, then a waterproof apron and eye protection such as a safety visor or goggles should be worn. Where there is a risk of scratching or biting, then stout protective gloves should be worn but consideration should be given to the impairment of dexterity, which may in itself present a hazard. If there is an aerosol risk then either the procedure causing the risk should be conducted in a safety cabinet or, if this is impossible, a full-face respirator or ventilated half-suit should be worn. With high category infectious agents it will be necessary for staff to undertake a complete change of clothing and to shower out on completion of their duties.

Safe practices

Safe practices during experimental procedures will help to reduce the hazards mentioned in previous sections. To avoid needlestick injury, all work involving infectious agents should be carried out only by staff experienced in inoculation techniques and in handling animals. When infecting or taking samples from animals they should be firmly restrained or anaesthetized. A variety of restraining devices is available to enclose an animal whilst leaving appropriate access for needles or catheters. However, the suitability of such equipment should be assessed before use.

When the contents of a sealed container are being withdrawn, the risk of contamination of the environment is dramatically reduced if the needle and the bottle cap are wrapped in cotton wool soaked in disinfectant during the withdrawal process. The risk is reduced almost to zero if, in addition, no air is pushed into the bottle before the fluid is withdrawn. When adjusting the volume of inoculum or removing air bubbles the needle should be inserted into a universal bottle containing cotton wool soaked in 70% alcohol or other suitable disinfectant. Dry cotton wool is not suitable as fluids spread very rapidly through it.

During inoculation the risk of contamination by leakage from the injection site is reduced if the skin surface is swabbed with disinfectant both before and after the procedure. The needle point should be directed away from the hands of both the operator and the restrainer and gloves should be worn by both operator and assistant in case of leakage of material.

During procedures involving sharp instruments, e.g. scalpels, only one person should be using them at any one time and they should never be passed from hand to hand. It is good practice to use disposable scalpels with integral blades rather than using the removable variety.

Accidental discharge of material by disconnection of the needle from the syringe can be prevented by the use of luer lock syringes. In this system, the needle locks on to the butt and is particularly recommended for inoculation procedures requiring considerable pressure, such as intradermal injections. Alternatively, suitably sized all-in-one syringes and needles may be available.

For most procedures it is safest to use disposable syringes and to discard the needle and syringe as one unit into a rigid plastic sharps container. This can then be sealed and removed for disposal by incineration or autoclaving followed by incineration. The use of evacuated containers for taking blood samples obviates the need to transfer blood into a second container and removes the necessity to take off the needle to prevent haemolysis during discharge of the contents. For the collection of other samples such as body fluids or inoculations that do not involve direct injection, the use of plastic catheters or pipettes is recommended.

The experimental infection of animals by aerosols poses hazards of infection to the operator. To reduce the risk, the entire procedure should be conducted inside a Class I or Class III biological safety cabinet. A Class I cabinet is an open-fronted, ventilated cabinet for personal protection with an unrecirculated inward airflow away from the operator. It is fitted with a HEPA filter to protect the environment from discharge of microorganisms. A Class III cabinet is a totally enclosed, ventilated cabinet which is gas-tight and is maintained under negative pressure. Both the supply and exhaust air are HEPA-filtered. Work is performed using attached glove sleeves and materials are removed or introduced via an attached pass box or dunk tank. Only the minimum amount of the animal's body, i.e. its snout, should be exposed to the aerosol and the apparatus should seal sufficiently well to the snout of the animal to prevent contamination of other parts of the body. After infection, the animals should be considered as a potential hazard and are best housed in some form of ventilated containment kept under negative pressure, such as a rigid cabinet or flexible film isolator. During the process of infection and for some time after, whilst the animal may be considered infectious, it may be necessary for staff to wear some form of respiratory protection. The aerosol risk associated with intranasal installation of an infectious organism makes this another procedure that is best carried out within a safety cabinet.

The risks associated with animal husbandry can be reduced in a number of ways. The use of clear plastic water bottles instead of glass will reduce the risk of cuts from broken glass and aerosols from dissemination of the contents of a dropped bottle. Where a high-category pathogen is being used, it is advisable to house animals on corn cob bedding or in gridded cages with absorbent disposable sheets instead of the conventional bedding materials such as sawdust. This will greatly reduce the hazard of inhaled bedding dust and will reduce the likelihood of filter blockage.

Disinfectants

Disinfectants have a number of uses in the animal facility. Their main function is to reduce contamination whilst a room or suite is in operation or whilst materials are in transit pending more rigorous treatment to make them safe, e.g. fumigation of a room at the end of an experiment or the autoclaving or incineration of waste materials. Disinfectants thus have an important role in good management and are used in footbaths, for wiping down work surfaces, for decontaminating contained samples which need to be removed from the facility and for swabbing skin surfaces during experimental procedures.

The disinfectant of choice will depend upon the microorganism under study, the amount of organic material present, and any potential harmful effects to equipment, personnel or animals present in the room (*see* Table 3). Whichever disinfectant is chosen, its effective dilution, the frequency with which fresh solutions need to be made and its compatibility with other chemicals should be incorporated in a code of practice.

Hypochlorites are effective against many bacteria, viruses, fungi and some spores. They are compatible with ionic and non-ionic detergents but may corrode metals and damage rubber, and are readily inactivated by organic material. Working solutions need to be changed frequently. Commonly used dilutions are 1000 ppm for general surface wiping, 2500 ppm for discard containers and 10,000 ppm for spillages. Hypochlorites sold in the UK contain 100,000 ppm available chlorine so dilutions of 1 : 100, 1 : 40 or 1 : 10 respectively need to be made for working solutions. More convenient are tablets such as 'Presept' which give a defined concentration of hypochlorite solution when added to a standard volume of water. Solid granules are available to soak up spillages and have the advantage of a longer storage life.

Table 3 Activities of some disinfectants

	Vegetative bacteria	Mycobacteria	Spores	Fungi	Lipid viruses	Non-lipid viruses	Inactivation by organic material
Hypochlorites	++	+	+	+	+	+	Serious
Phenolics	++	++	−	++	+	Variable	Slight
Alcohols	++	+	−	−	+	+	Moderate
Aldehydes	++	++	++	++	+	+	Slight
QACs	++	−	−	+	−	−	Serious
Iodophors	++	++	+	++	+	+	Serious

Key: ++Good. +Fair. − Ineffective

Clear soluble phenolics are based on xylenols and most contain either soap or detergent. They are compatible with ionic and non-ionic detergents and metals. They are effective against bacteria, including mycobacteria, fungi and lipid-containing viruses but are inactive against spores. Activity is not greatly reduced by the presence of organic material but it is recommended that they should be made up to the manufacturer's dilution for 'dirty conditions'. Diluted phenolics remain active for 24 hours.

Alcohols are effective against many bacteria and lipid-containing viruses, and are quick-acting, stable and compatible with metals. They have poor penetration of organic materials and are flammable. The most effective strength is 70–80% v/v solution in water.

Aldehydes are usually used either as a gaseous fumigant (formaldehyde) or as a liquid disinfectant (glutaraldehyde). They are active against bacteria, spores, fungi and both lipid-containing and non-lipid containing viruses. Formaldehyde gas is too irritant for general use and is considered toxic with a MEL of 2 parts per million. It is most active above 20°C and requires humidity above 70% RH to be effective. Its main use is in the fumigation of rooms or safety cabinets. Formaldehyde gas for space decontamination may be produced either by boiling a formalin solution ($100\,cm^3$ formalin + $900\,cm^3$ water per $28.3\,m^3$ ($1000\,ft^3$) of space, or by heating solid paraformaldehyde at a concentration of $10.5\,g/m^3$ ($0.3\,g/ft^3$).

Glutaraldehyde usually requires an activator which raises and buffers the pH of the solution. Once activated, the solution will remain effective for 14–28 days. It is useful for disinfecting equipment or surfaces where corrosion may be a problem. Glutaraldehyde vapour is toxic, irritant and can lead to respiratory sensitivity.

Quaternary ammonium compounds (QACs) are cationic detergents which are effective against bacteria, lipid-containing viruses and some fungi. They have the advantage of being stable and non-toxic but are inactivated by organic matter and soaps.

Iodophors are effective against bacteria, including mycobacteria, spores, fungi and both lipid-containing and non-lipid-containing viruses. They are inactivated by organic material and tend to stain skin and surfaces. For general disinfection, they should be diluted to give 75–150 ppm iodine.

Monitoring

It is a wise precaution to screen animals for the presence of naturally occuring pathogenic organisms on a regular basis. Animals obtained from accredited breeders are usually fully documented as to their health history; nevertheless, it is good practice to isolate newly delivered batches of animals for a defined 'quarantine' period. Animals bred in-house should be tested by culturing for bacterial and fungal contamination, by direct microscopy or stained smears for protozoal and higher parasites and by serological screening for viruses. Recommendations for the health monitoring of mouse, rat, hamster, guineapig and rabbit breeding colonies have been made by a FELASA working group and have been published in *Laboratory Animals* (FELASA 1994). There are several commercial laboratories offering a comprehensive service if this cannot be carried out within the facility. Sentinel animals known to be free of viruses, parasites and pathogenic bacteria are often used to detect subclinical infections, particularly in the case of mice where suitably susceptible animals are readily obtainable. Sentinel animals are also useful when monitoring immunosuppressed animals which may themselves show no immune response to a viral infection. Sentinels should remain in place for a period of up to 10 weeks or until they show signs of infection, i.e. clinical signs or development of specific antibodies.

Testing the environment for contamination can be performed by the use of settle plates, direct contact plates, swabbing and attempting to culture in broths or by air samplers. Settle plates are the simplest method for the assessment of aerial contamination. A number of petri dishes containing appropriate selective or non-selective media are exposed media side up for specified periods and are subsequently incubated, colonies counted and organisms identified. Contact plates are small petri dishes filled to the brim with agar media. They are inverted over and gently pressed on the surface to be sampled, the lids replaced and the plates incubated. Contact samplers are also available in slide form. Alternatively, surfaces may be swabbed using cotton wool or alginate swabs which are then wiped across agar media or broken off into broths and incubated.

For sampling large volumes of air, various air samplers or impingers are available. The efficacy of these has been reviewed by Hambleton *et al.* (1992).

The efficacy of a fumigation process can be tested by positioning bacterial spore strips at various strategic points in a room, cabinet or isolator and attempting to grow the indicator organism on completion of the fumigation process.

Risk assessment

When conducting any experiment involving pathogenic micro-organisms or working with naturally infected animals, an assessment of the risks involved must be undertaken. From this assessment a comprehensive code of practice can be drawn up and in the UK a COSHH (Control of Substances Hazardous to Health) form or other similar statement of the risks involved must be completed. All staff involved in a programme of work should be made aware of any associated risk assessments and codes of practice or protocols, and should sign to say that they have read and understood these documents.

The initial assessment must take into account the category of pathogen: in the UK microorganisms have been categorized by the Advisory Committee on Dangerous Pathogens (ACDP) as described earlier in this chapter. If the experiment involves genetically modified microbes then the local genetic modification committee must be consulted and the level of containment assessed according to Genetic Modification Advisory Group (GMAG) guidelines. Other factors that need to be taken into account for a local code of practice include the volume of inoculum, the route of infection in the animal, the likely route of infection in man, the expected route and level of excretion from the animal of the organism and its likely survival in aerosols, blood, faeces or other body fluids.

For more detailed information on regulations regarding genetic modification and laws regarding safety, see other chapters in this handbook. For a summary of ACDP containment requirements for animals, see Table 4 below. The subject of containment of infected animals is considered in the ACDP publication (1997), *Working safely with research animals: Management of infection risks*.

In summary, the use of infectious agents in experimental animals still presents a variety of hazards which must not be ignored. With good management, appropriate equipment and facilities, adequate training and the use of high-quality animals, these risks can be reduced to a minimum and results of high scientific integrity can be obtained.

A checklist for conducting a risk assessment should contain the following points:

- Does the work need to be conducted in animals?
- The hazard category of the organism and the level of containment required.
- The concentration and volume of the inoculum.

Table 4 ACDP containment levels for animals (a summary)

Requirement	Containment level 1	2	3	4
Room sealable for fumigation	No	No	Yes	Yes
Ventilation				
Inward airflow	No	Optional	Yes	Yes
Mechanical	No	Optional	Yes	Yes
Mechanical independent	No	No	No	Yes
HEPA filtration	No	No	Single-extract	Single-input/ Double-extract
Double door entry	No	No	Yes	Yes
Airlock	No	No	No	Yes
Handwashing facility	Yes	Yes	Yes	Shower
Shower	No	No	Available	Yes
Effluent treatment	No	No	No	Yes
Autoclave				
On site	Yes	Yes	No	No
In suite	No	No	Yes	No
Integral double ended	No	No	No	Yes
Biological safety cabinet	No	Class I	Class I or III	Class III
Change of clothing	Yes	Yes	Yes, disinfect after use	Yes, autoclave after use
Incinerator on site	No	Optional	Yes	Yes
Limited access	Yes	Yes	Yes	Yes
Insect and rodent control	No	Yes	Yes	Yes
Containment of animals or use RPE	No	No	Yes	Yes

- The route of challenge and the risks associated with the challenge procedure: are staff proficient in these procedures?
- The likelihood of the organism being excreted after challenge.
- The risks associated with day-to-day husbandry.
- The risks associated with any sampling procedure.
- How samples are to be transported and processed.
- What samples are to be taken *post mortem*?
- How are the processes of challenge, husbandry, sampling and necropsy to be safely contained to protect personnel?
- Are vaccines available and advisable?
- How are waste materials and carcasses to be disposed of in a safe manner?
- How can the room and ancillary equipment be made safe at the end of the experiment?

In addition to this all staff associated with the work should be made aware, in the form of written instructions, of what procedures are to be followed in the event of accidental release of the agent, and of what to do in the event of being bitten or scratched by the animal or in the event of a cut or needlestick injury.

A Code of Practice should be in place which defines how such incidents are to be reported and investigated, taking due consideration of the requirements of The Reporting of Incidents, Diseases and Dangerous Occurrences Regulations 1995 (RIDDOR) (*see* Chapters 7 and 8).

Acknowledgement

Finally I must express my thanks to the late Dr E A Boulter, the author of this chapter in the forerunner of this Handbook (*Safety in the Animal House*, 1981), whose authoritative work has been an invaluable guide to this update.

References

Advisory Committee on Dangerous Pathogens (1994) *Precautions for work with human and animal transmissible spongiform encephalopathies*. London: HMSO

Advisory Committee on Dangerous Pathogens (1995) *Categorisation of biological agents according to hazard and categories of containment*. 4th edn. Sudbury: HSE Books

Advisory Committee on Dangerous Pathogens (1997) *Working safely with research animals: Management of infection risks*. Sudbury: HSE Books

Advisory Committee on Dangerous Pathogens (1998) *Working safely with primates: Management of infection risks*. Sudbury: HSE Books

Advisory Committee on the Microbiological Safety of Food (1995) *Report on verocytotoxin-producing Escherichia coli*. London: HMSO

Anusz Z, Szweda W, Popko J, Trybala E (1992) Is Aujeszky's disease a zoonosis? *Przeglad Epidemiologizny* **46**, 181–6

Arashima Y, Kumasaka K, Kawano K, Asano R, Hokari S, Murasugi E (1992) Studies on giardiasis as a zoonosis. III. Prevalence of Giardia among dogs and owners in Japan. *Kansenshogaku-Zasshi. Journal of the Japanese Association for Infectious Disease* **66**, 1062–6

Arita I, Jesek Z, Khodakevich L, Kalasi R (1985) Human monkey pox: a newly emerged orthopox virus zoonosis in tropical rainforests of Africa. *American Journal of Tropical Medicine and Hygiene* **34**, 781–9

Asai T, Kinjo T, Minamoto N, Sugiyama M, Matsubayashi N, Narama I (1991) Prevalence of antibodies to five selected zoonosis agents in monkeys. *Journal of Veterinary Medical Science* **53**, 553–9

Baron RC, McCormick JB, Zubeir OA (1983) Ebola virus disease in Southern Sudan: hospital dissemination and intrafamilial spread. *Bulletin of the World Health Organization* **61**, 997–1003

Baskerville A, Humphrey TJ, Fitzgeorge RB, Cook RW, Chart H, Rowe B, Whitehead A (1992) Airborne infection of laying hens with *Salmonella enteritidis* phage type 4. *Veterinary Record* **130**, 395–8

Becker S, Feldman H, Will C, Slenczka W (1992) Evidence for occurrence of filovirus antibodies in humans and imported monkeys: do subclinical filovirus infections occur world wide? *Medical Microbiology and Immunology, Berlin* **181**, 43–55

Biggar RJ, Schmidt TJ, Woodall JP (1977) Lymphocytic choriomeningitis in laboratory personnel exposed to hamsters inadvertently infected with LCM virus. *Journal of the American Veterinary Medical Association* **171**, 829–32

Bignall J (1995) Hantaviruses: the rodents take revenge. *Lancet* **345**, 1564

Blackwell JM (1988) Bacterial infections. In: *Genetics of resistance to bacterial and parasitic infections* (Wakelin DM, Blackwell JM, eds). London: Taylor and Francis: 63–101

Blaser M, LaForce FM, Wilson NA, Wang WLL (1980) Reservoirs for human campylobacteriosis. *Journal of Infectious Diseases* **141**, 665–9

Boulter EA (1975) The isolation of monkey B virus (*Herpesvirus simiae*) from the trigeminal ganglia of a healthy seropositive rhesus monkey. *Journal of Biological Standardization* **3**, 279–80

Boulter E, Grant DP (1977) Latent infection of monkeys with B virus and prophylactic studies in a rabbit model of this disease. *Journal of Antimicrobial Chemotherapy* **3**, 107–13

Breman JG, Kalisa R, Steniowski MV, Zanotto E, Gromyko AI, Arita I (1980) Human monkeypox 1970–1979. *Bulletin of the World Health Organization* **58**, 165–82

Bruins SC, Tight RR (1979) Laboratory acquired gonococcal conjunctivitis. *Journal of the American Medical Association* **241**, 274

Chevalier X, Martigny J, Avouac B, Larget-Piet B (1991) Report of 4 cases of *Pasteurella multocida* septic arthritis. *Journal of Rheumatology* **18**, 1890–2

Collins CH (1993) *Laboratory acquired infections.* Oxford: Butterworth-Heinemann: 17

Collins CH, Kennedy DK (1987) Microbiological hazards of occupational needlestick and 'sharps' injuries. *Journal of Applied Bacteriology* **62**, 385–402

Dammeijer PF, McCombe AW (1991) Meningitis from canine *Pasteurella multocida* following mastoidectomy. *Journal of Laryngology and Otology* **105**, 571–2

Daniel MD, Letvin N., King NW, Kannagi M, Sehgal PK, Hunt RD, Kanki PJ, Essex M, Desrosiers RC (1985) Serological identification and characterisation of a macaque T-lymphotropic retrovirus closely related to HTLV-III. *Science* **228**, 1199–204

Darlow HM (1972) Safety in the microbiology laboratory: an introduction. In: *Safety in microbiology* (Shapton DA, Board RG, eds). London: Academic Press: 1–20

Dennis MJ (1986) The effects of temperature and humidity on some animal diseases – a review. *British Veterinary Journal* **142**, 472–85

Desmyter J (1983) Laboratory rat-associated outbreak of haemorrhagic fever with renal syndrome due to Hantaan-like virus in Belgium. *Lancet* **ii**, 1445–8

Dietzman DE, Fuccillo DA, West FJ, Moder F, Sever JL (1973) Conjunctivitis associated with Coxsackie B1 virus in a laboratory worker. *American Journal of Ophthalmology* **75**, 1045–6

Dubey JP (1993) Intestinal protozoa infections. *Veterinary Clinical North American Small Animal Practice* **23**, 37–55

FELASA (1994) Recommendations for the health monitoring of mouse, rat, hamster, guineapig and rabbit breeding colonies. *Laboratory Animals* **28**, 1–12

Field PR, Moyle CG, Parnell PM (1972) The accidental infection of a laboratory worker with *Toxoplasma gondii. Medical Journal of Australia* **ii**, 196–8

Fisher-Hoch SP, Perez-Oronoz GI, Jackson EL, Hermann LM, Brown BG (1992) Filovirus clearance in non-human primates. *Lancet* **340**, 451–3

Fordham JN, McKay-Ferguson E, Davies A, Blyth T (1992) Rat bite fever without the bite. *Annals of Rheumatic Disease* **51**, 411–12

Fox JG, Claps MC, Taylor NS, Maxwell KO, Ackerman JI, Hoffman SB (1988) *Campylobacter jejuni/coli* in commercially reared beagles: prevalence and serotypes. *Laboratory Animal Science* **38**, 262–5

Fraumeni JF, Stark CR, Gold E, Lepow ML (1970) Simian virus 40 in polio vaccine; follow-up of newborn recipients. *Science* **167**, 59–60

Gardner M, Marx P, Maul D, Osborn K, Hendrickson R, Lerche N, Bencken B, Bryant M (1984) Simian AIDS: Evidence for a retroviral aetiology. *Haematological Oncology* **2**, 259–68

Gordon Smith CE, Simpson DIH, Bowen ETW, Zlotnik I (1967) Fatal human disease from vervet monkeys. *Lancet* **ii**, 1119

Gross L (1983) *Oncogenic viruses*. Oxford: Pergamon Press: 980–1002

Grubb R, Midtvedt T, Norin E (1988) *The regulatory and protective role of the normal microflora*. Basingstoke: Macmillan Press

Guion TL, Sculco TP (1992) *Pasteurella multocida* infection in total knee arthroplasty. Case report and literature review. *Journal of Arthroplasty* **7**, 157–60

Gupta H, Tandon BN, Sriramachari S, Joshi YK, Iyenger B (1990) Animal transmission of enteric non-A, non-B hepatitis infection in *Macaca mulatta* by faeco-oral route. *Indian Journal of Medical Research* **91**, 87–90

Gustafson DA, Moses HE (1951) Isolation of Newcastle disease virus from the eye of a human being. *Journal of the American Veterinary Medical Association* **118**, 1–2

Hadley KM, Carrington D, Frew CE, Gibson AA, Hislop WS (1992) Ovine chlamydiosis in an abattoir worker. *Journal of Infection* **25**(Suppl. 1), 105–9

Hammadeh GN, Turner BW, Trible W, Hoffman BJ, Anderson RM (1992) Laboratory outbreak of Q fever. *Journal of Family Practitioners* **35**, 683–5

Hambleton P, Bennett AM, Leaver G (1992) Biosafety monitoring devices for biotechnology processes. *Trends in Biotechnology* **10**, 192–9

Hanel E, Alg RL (1955) Biological hazards of common laboatory procedures. II. The hypodermic syringe and needle. *American Journal of Medical Technology* **21**, 343–6

Hayami M, Ido E, Miura T (1994) Survey of simian immunodeficiency virus among nonhuman primate populations. In: *Simian immunodeficiency virus* (Letvin NL, Desrosiers RC, eds). Berlin: Springer-Verlag: 1–20

Hayami M, Komuro A, Nozawa K (1984) Prevalence of antibody to adult T-cell leukaemia virus-associated antigens (ATLA) in Japanese monkeys and other non-human primates. *International Journal of Cancer* **33**, 179–83

Henderson DW (1952) An apparatus for the study of airborne infection. *Journal of Hygiene, Cambridge* **50**, 53–68

Hennessen W (1969) Epidemiology of Marburg virus disease. In: *Hazards of handling simians*. Laboratory Animal Handbooks No. 4 (Perkins FT, O'Donoghue PN, eds). London: Laboratory Animals Ltd: 137–42

Henrickson R, Maul DH, Osborn KG, Sever JL, Madden DL, Ellingsworth LR, Anderson JH, Lowenstine LJ, Gardner MB (1983) Epidemic of acquired immunodeficiency syndrome in rhesus monkeys. *Lancet* **i**, 388–90

Jesek Z, Grab B, Szczeniowski M, Paluki KM, Mutombo M (1988) Clinico-epidemiological features of monkeypox patients with an animal or human source of infection. *Bulletin of the World Health Organization* **66**, 459–64

Kalter SS, Heberling RL, Cooke AW, Barry JD, Tian PY, Northam WJ (1997) Viral infections of nonhuman primates. *Laboratory Animal Science* **47**, 461–7

Kaspareit-Rittinghausen J, Wullenweber M, Deerberg F, Farouq M (1990) Pathological changes in *Streptobacillus moniliformis* infection in C57BL/6J mice. *Berliner und Munchener Tierarztliche-Wochenschrift* **103**, 84–7

Kennedy FM, Astbury J, Needham JR, Cheasty T (1993) Shigellosis due to occupational contact with non-human primates. *Epidemiology and Infection* **110**, 247–51

Khabbaz RF, Rowe T, Murphey-Corb M, Heneine WM, Schable CA, George JR, Pau CP, Parekh BS, Lairmore MD, Curran JW, Kaplan JE, Schochetman G, Folks TM (1992) Simian immunodeficiency virus needlestick accident in a laboratory worker. *Lancet* **340**, 271–3

Kirchner BK, Lake SG, Wightman SR (1992) Isolation of *Streptobacillus moniliformis* from a guinea pig with granulomatous pneumonia. *Laboratory Animal Science* **42**, 519–21

Kulagin SM, Fedorova NI, Ketiladze ES (1962) Laboratory outbreak of haemorrhagic fever with renal syndrome. *Zhurnal Mikrobiologie, Epidemiologie i Immunologie* **33**, 121–6

Kumar A, Kannampuzha P (1992) Septic arthritis due to *Pasteurella multocida*. *Southern Medical Journal* **85**, 329–30

Lapsin BA, Shevtsova ZV (1990) Persistence of spontaneous and experimental hepatitis A in rhesus macaques. *Experimental Pathology* **39**, 59–60

Le Guenno B, Formentry P, Wyers M, Gounon P, Walker F, Boesch C (1995) Isolation and partial characterisation of a new strain of Ebola virus. *Lancet* **i**, 1330–2

Letvin NL, Eaton KA, Aldrich WR, Sehgal PK, Blake BJ, Schlossman SF, King NW, Hunt RD (1983) Acquired immunodeficiency syndrome in a colony of macaque monkeys. *Proceedings of the National Academy of Science, USA* **80**, 2718–22

Levy BM, Mirkovic RR (1971) An epizootic of measles in a marmoset colony. *Laboratory Animal Science* **21**, 33–9

Lillie LE, Thompson RG (1972) The pulmonary clearance of bacteria by calves and mice. *Canadian Journal of Comparative Medicine* **36**, 129–37

Lloyd G, Jones N (1986) Infection of laboratory workers with Hantaan virus acquired from immunocytomas propagated in laboratory rats. *Journal of Infection* **12**, 117–25

Mahy BW, Dykewicz C, Fisher-Hoch S, Ostroff S, Tipple M, Sanchez A (1991) Viral zoonoses and their potential for contamination of cell cultures. *Development of Biological Standards* **75**, 183–9

Martini GA, Siegert K (1971) *Marburg virus disease*. Berlin: Springer-Verlag

Mayr B (1993) Pets as permanent excretors of zoonoses pathogens. *Zentralblatt fur Hygiene und Umweltmedizin* **194**, 214–22

McKenna P, van der Groen G, Hoofd G, Beelaert G, Leirs H, Verbagen R, Kints JP, Cormont F, Nisol F, Bazin H (1992) Eradication of hantavirus infection among laboratory rats by application of Caesarian section and a foster mother technique. *Journal of Infection* **25**, 181–90

Medical Research Council Simian Virus Committee (1990) *The management of simians in relation to infectious hazards to staff*. London: Medical Research Council

Melendez LV (1968) Herpes T infection in man? *Laboratory Primate Newsletter* **7**, 1

Melnick JL, Curnen EC, Sabin AB (1948) Accidental laboratory infection with human dengue virus. *Proceedings of the Society for Experimental Biology and Medicine* **68**, 198–200

Meyer A, Esposito JJ, Gras F, Kolakowski T, Fatras M, Muller G (1991) First appearance of monkey pox in human beings in Gabon. *Medicine Tropicale (Marseilles)* **51**, 53–7

Miller CD, Songer JR, Sullivan JF (1987) A twenty five year review of laboratory acquired human infection at the National Animal Disease Center. *American Industrial Hygiene Association Journal* **48**, 271–5

Minton EJ (1990) *Pasteurella pneumotropica*: meningitis following a dog bite. *Postgraduate Medical Journal* **66**, 125–6

Montali RJ, Scanga CA, Pernikoff D, Wessner DR, Ward R, Holmes HV (1993) A common-source outbreak of callitrichid hepatitis in captive tamarins and marmosets. *Journal of Infectious Disease* **167**, 946–50

Morbidity and Mortality Weekly Report (MMWR) (1987) B virus infection in humans – Pensacola, Florida. *MMWR* **36**, 289–96

MMWR (1989) B virus infection in humans – Michigan. *MMWR* **38**, 453–4

MMWR (1990) Update: Filovirus infection in animal handlers. *MMWR* **39**, 221

MMWR (1991) Update – non-human primate importation. *MMWR* **40**, 684–5, 691

MMWR (1993) Tuberculosis in imported non-human primates, United States, June 1990 – May 1993. *MMWR* **42**, 572–6

Mortimer EA, Lepow ML, Gold E, Robbins FC, Burton GJ, Fraumeni JF (1981) Long term follow-up of persons inadvertently inoculated with SV 40 as neonates. *New England Journal of Medicine* **305**, 1517–18

Mukinda VBK, Mwema G, Kilundu M, Heymann DL, Khan AS, Esposito JJ and others (1997) Re-emergence of human monkeypox in Zaire in 1996. *Lancet* **349**, 1449–50

Nanda M, Curtis VT, Hilliard JK, Bernstein ND, Dix RD (1990) Ocular histopathologic findings in a case of human Herpes B virus infection. *Archives of Ophthalmology* **108**, 713–16

Palmer AE (1987) B-virus, *Herpesvirus simiae*: Historical perspective. *Journal of Medical Primatology* **16**, 99–130

Pike RM (1976) Laboratory associated infection: summary and analysis of 3,921 cases. *Laboratory Animal Science* **13**, 105–14

Pike RM, Sulkin SE, Schulze ML (1965) Continuing importance of laboratory acquired infection. *American Journal of Public Health* **55**, 190–9

Reitman M, Alg RL, Miller WS, Gross WH (1954) Potential hazards of laboratory techniques. III. Virus techniques. *Journal of Bacteriology* **68**, 548–54

Saxinger W, Blattner WA, Levine PH, Clark J, Biggar R, Hoh M, Moghissi J, Jacobs P, Wilson L, Jacobsen R (1984) Human T-cell leukaemia virus (HTLV-1) antibodies in Africa. *Science* **225**, 1473–6

Schlech WF (1988) The risk of infection in anaesthesia practice. *Canadian Journal of Anaesthesia* **35**, 346–51

Schmidt DR, Winn RE, Keef TJ (1989) Leptospirosis. Epidemiological features of a sporadic case. *Archives of Internal Medicine* **149**, 1878–80

Simpson DIH (1977) *Marburg and Ebola virus infections. A guide for their diagnosis, management and control.* World Health Organization Offset Publication No. 36. Geneva: WHO

Stableforth AW (1953) Animal brucellosis. In: *Advances in the control of zoonoses*. World Health Organization Monograph Series No. 19. Geneva: WHO: 71–87

Stephenson CB, Jacob JR, Montali RJ, Holmes KV, Muchmore E, Compans RW, Arms ED, Buchmeier MJ, Landford RE (1991) Isolation of an arenavirus from a marmoset with callitrichid hepatitis and its serological association with disease. *Journal of Virology* **65**, 3995–4000

Stiglmair-Herb MT, Scheid R, Hanichen T (1992) Spontaneous inclusion body hepatitis in young tamarins. I. Morphological study. *Laboratory Animals* **26**, 80–7

Sulkin SE, Pike RM (1951) Laboratory acquired infection. *Journal of the American Medical Association* **147**, 1740–5

Tabor E (1989) Non-human primate models for non-A, non-B hepatitis. *Cancer Detection and Prevention* **14**, 221–5

Tappero JW, Mohle-Boetani J, Koehler JE, Swaminathan B, Berger TG, LeBoit PE, Smith LL, Wenger JD, Pinner RW, Kemper CA (1993) The epidemiology of bacillary angiomatosis and bacillary peliosis. *Journal of the American Medical Association* **269**, 770–5

Teelman K, Weihe WH (1974) Microorganism counts and distribution patterns in air-conditioned animal laboratories. *Laboratory Animals* **8**, 109–18

Tjong JWR, van Zeijl JH (1991) Fatal complication after dog bites with *Capnocytophaga canimorsus. Nederlands Tijdschrift voor Geneeskdkunde* **135**, 138–40

Tontis A, Zwalen R (1991) Chlamydia infection in sheep and goats. With a reference to its significance as a zoonosis. *Tierarztliche Praxis* **19**, 617–23

Tribe GW (1969) Clinical aspects of the detection and control of tuberculosis in newly imported monkeys. In: *Hazards of handling simians*. Laboratory

Animal Handbooks No. 4 (Perkins FT, O'Donoghue PN, eds). London: Laboratory Animals Limited: 19–23

Tuffery AA, Innes JRM (1963) Diseases of laboratory mice and rats. In: *Animals for research* (Lane-Petter W, ed). London: Academic Press: Chapter 3

Umenai T, Lee PW, Toyoda T, Yoshinaga K, Horiuchi T, Lee HW, Saito T, Hongo M, Nokunaga T, Ishida N (1979) Korean haemorrhagic fever in staff in an animal laboratory. *Lancet* **i**, 1314–16

Vizoso AD (1975) Recovery of *Herpes simiae* (B virus) from both primary and latent infections in Rhesus monkeys. *British Journal of Experimental Pathology* **56**, 485–8

Weber DJ, Hansen AR (1991) Infections resulting from animal bites. *Infectious Disease Clinics of North America* **5**, 663–80

Weihe WH (1975) Phase maps for particles and microorganisms in animal quarters. *Laboratory Animals* **9**, 353–65

Welcker A (1938) Laboratoriumsinfektionen mit Weilscher Krankheit. *Zentralblatt fur Bakteriologie* **141**, 400–10

Werth D (1989) The occurrence and significance of *Chlamydia psittaci* and *Coxiella burnetii* in dogs and cats. A study of the literature. *Berliner und Munchener Tierarztliche Wochenschrift* **102**, 156–61

World Health Organization (1978) Ebola haemorrhagic fever in Sudan, 1976 – report of a WHO/International Study Team. *WHO Bulletin* **56**, 247–70

Wright LJ, Barker LF, Mickenberg ID, Sheldon MW (1968) Laboratory acquired typhus fevers. *Annals of Internal Medicine* **63**, 731–8

Wullenweber M, Kaspareit-Rittinghausen T, Farouq M (1990) *Streptobacillus moniliformis* epizootic in barrier maintained C57BL/6J mice and susceptibility to infection of different strains of mice. *Laboratory Animal Science* **40**, 608–12

Zamiatina NA, Andzhaparidze AG, Balaian MS, Sohol AV, Titova IP, Karetnyi IUV, Poleschuk VF (1990) Development of an infection in monkeys as a result of their sequential natural and experimental exposure to the hepatitis A virus. *Voprosy Virusologii* **35**, 122–5

Further reading

Bell JC, Palmer SR, Payne JM (1988) *The zoonoses: infections transmitted from animals to man*. London: Edward Arnold

Brack M (1987) *Agents transmissible from simians to man*. Berlin: Springer-Verlag

Brown DW (1993) Filoviruses and imported non-human primates. *PHLS Microbiology Digest* **10**, 195–8

Collins CH (1993) *Laboratory acquired infections*. Oxford: Butterworth-Heinemann

Health and Safety Commission (HSC) (1992) Education Services Advisory Committee *Health and safety in animal facilities*. London: HMSO

HSC Health Services Advisory Committee (1991) *Safe working and the prevention of infection in the mortuary and post mortem room*. London: HMSO

Journal of Medical Primatology (1987) Special Issue. Biohazards associated with natural and experimental diseases of non-human primates. **16** (2)

Owen DG (1992) *Parasites of laboratory animals*. Laboratory Animal Handbooks No. 12. London: Royal Society of Medicine Press on behalf of Laboratory Animals Ltd

Perkins FT, O'Donoghue PN, eds (1981) *Hazards of handling simians*. Laboratory Animal Handbooks No. 4. London: Laboratory Animals Ltd

Poole T, ed (1999) *UFAW Handbook on the care and management of laboratory animals*. Vols 1&2. 7th edn. Oxford: Blackwell Science

Quinn PJ, Carter ME, Markey B, Carter GR (1994) *Clinical veterinary microbiology*. London: Wolfe

World Health Organisation. (1993) *Laboratory biosafety manual*. Geneva: WHO

4

Genetically modified (transgenic) animals

M W Smith

Contents

Introduction

In many animal units it is possible that the only requirement for facilities for transgenic animals may be for the housing or breeding of established strains. This might be considered as constituting little real risk to the staff apart from that to be expected from any other group of animals. Hazards must nevertheless be considered and allowed for.

Some animal units may, however, have inbuilt or planned facilities for the production of new transgenic strains and, depending on the techniques used, will be considered as presenting a variable level of potential hazard. The concept of risk relates partly to the use of techniques involving the handling of DNA and also viruses or other organisms as potential vectors and partly to the public perception of the possibly unknown and indefinable results of genetic modification in itself. Strong public feeling has been expressed concerning the accidental or deliberate release to the wild of genetically modified organisms (including animals) which could have an as yet unknown effect on the environment.

This has been accepted by the Advisory Committee on Genetic Modification which recognizes that accidental introduction to the environment may cause problems relating to the character of the animals themselves (as has happened in the past following the release of imported species such as mink). Problems could also arise as the result of the transmission of modified genes to domestic or wild animals. Physical or behavioural changes in transgenic animals could have a variety of implications for the environment, e.g. due to increases in size, food intake, or reproductive capacity. There could be excessive growth in numbers or competition with wild populations.

It is possible that some of those working in this field believe that attempting to predict such hazards is an impossible exercise, which hinders academic innovation. They may also feel that basic research implies working on the fringes of risk for which there is an unforeseen and unpredictable price to pay. However, it is considered that a precautionary approach is reasonable and necessary and this is expressed in two sets of regulations introduced in response to EC Directives 90/219 and 90/220. These are:

a) The Genetically Modified Organisms (Contained Use) Regulations (1992), as amended by the Genetically Modified Organisms (Contained use) (Amendment) Regulations 1996 & 1998, made under the powers of the Health and Safety at Work etc. Act (1974) and the European Communities Act (1972).

b) The Genetically Modified Organisms (Deliberate Release) Regulations (1992) (as revised in 1995), introduced under the Environmental Protection Act (1990).

Since this chapter largely deals with the safety aspects of the production and use of transgenic animals in the research environment, only the first of these regulations will be dealt with in any detail.

What are transgenic animals?

Animals showing naturally occurring characteristics have been selectively bred for many years. This is in effect a modification of their DNA, the results of which are self-evident in the great variety of breeds of companion and farm animals currently available. The use of recombinant DNA techniques has, however, made it possible to speed up this process by isolating specified genetic material and transferring it to other animals. The term 'transgenic' was first used in relation to mice (Gordon & Ruddle 1981) and could be defined as the transmission of foreign DNA from animal or plant sources to the animal genome at the embryonic stage of development. Genetically modified animals which may result from this process are called transgenic animals. This description has since been extended to include animals produced by genetic modification using gene targeting or gene deletion techniques.

A number of small animal species have been used in the production of transgenic animals, notably the mouse. However other animals such as rabbits, pigs and sheep (Hammer *et al.* 1985) and fish (Chen & Powers 1990) are of increasing interest.

How are transgenic animals made?

The level of potential risk involved in producing transgenic animals is influenced by such factors as the source of any foreign DNA, the method of producing the transgenic construct and the effect of the procedure on the embryo itself. A number of methods are in use and these in general outline include:

(a) Microinjection of foreign DNA into one pronucleus (usually the male) of the fertilized egg

In this method, embryos are recovered from the female reproductive tract the day after mating. After cleaning they are transferred to a micromanipulator microscope. At this stage the two gametes have fused, but the nuclei (called pronuclei) are still separate. A solution of the selected foreign DNA is then injected usually into the larger (male) pronucleus through a specially made needle. To some extent the choice of which pronucleus is used depends on position and visibility and on the skill of the operator. Treated embryos are later transferred surgically to surrogate mothers and continue thereafter to develop until normal parturition (Hogan *et al.* 1986).

(b) Transfection via the use of embryonic stem cells

Embryonic stem (ES) cells can differentiate into many cell types and are termed pluripotent. They can be cultured from the inner cell mass of a blastocyst (Evans & Kaufman 1981). After specific treatments they can then be transferred back into other blastocysts by micro-injection into the blastocoele. Under appropriate circumstances the introduced ES cells can contribute to all the embryonic tissues including the germline, and it is thus possible to modify or target specific genes in the ES cells and then introduce the modified genes into selected blastocysts. In this way DNA can be modified or replaced by selected genetic material and this process is known as gene targeting (Thomas *et al.* 1986).

(c) The use of viral vectors

Retroviruses (some of which are oncogenic) are single-stranded RNA viruses which replicate by means of a DNA provirus integrated into cellular DNA. Over 20 years ago it was shown (Jaenish 1976) that when early mouse embryos are infected with retroviruses, the integrated proviral DNA can be found in the cells of the resultant adult animals. Such animals can also transmit the proviral DNA to their offspring. Methods have since been developed to substitute heterologous genetic material for portions of the retrovirus, and a wide range of viral vectors has been developed (Valerio 1992).

One method of using vector viruses has been to culture embryos on a monolayer of virus-producing cells, or alternatively the virus may be injected by micropipette into the blastocoele after preliminary treatment to include the required DNA.

The use of viral vectors in the production of transgenic animals has decreased markedly with the advent of other methods. Similarly, the use of viruses which may shed in transgenic animals is now very rare.

Further developments in the methods of production of transgenic animals include:

1) The co-culturing of embryos on monolayers of selected cultured ES cells.
2) The use of yeast artificial chromosomes (YACs) as a means of handling larger gene fragments (Burke *et al.* 1987, Pachnis *et al.* 1990).
3) The use of gene ablation as a means of destroying all cells expressing a particular gene (Breitmen *et al.* 1987).

Regulations concerning the generation and use of transgenic animals

The implementation of the EC Directive 90/219 was followed by the introduction in the UK of the current 'Contained Use' regulations mentioned earlier in this chapter, which are administered by the Health and Safety Executive (HSE). These regulations, which came into effect in 1993 and were amended in 1996 and 1998,

a) require a risk assessment to be carried out before work involving genetic modification begins, based on defined methods;
b) require certain work with genetically modified organisms (GMOs) to be allocated a suitable level of containment. Such work includes the production, breeding and scientific use of transgenic animals;
c) require the person(s) involved to notify the HSE of the intention to use premises for genetic modification work for the first time;
d) require the formation of a local genetic modification safety committee to discuss and advise on work to be introduced.

Classification of work involving genetically modified animals

This classification is included for purposes of completeness only, since type A and B operations apply only to work with microorganisms. However, since microorganisms may be used in the production of transgenic strains, it is useful to be aware of these sections.

Work

Work is divided into two main groups:

Type A operations. These include the use of genetically modified organisms for teaching, research, or development and for small-scale non-industrial operations, whose standards reflect good microbiological practice. Also, organisms should be capable of being rendered inactive by standard methods.

Type B operations. These include all other methods and procedures including industrial processes.

Organisms

Organisms themselves are also classified according to four main criteria:

a) The nature of the host organism.
b) The nature of the cloned DNA.
c) The nature of the cloning vector.
d) The nature of the resulting genetically modified organism as a whole.

Two main groups are defined:

Group I. This includes any organisms considered unlikely to cause disease to humans, animals or plants, or unlikely to cause harm to the environment.

Group II. This includes any organisms that do not meet these criteria and include human, animal or plant pathogens.

Such groupings are not absolute, in that a Group I host may not be classified as a Group I genetically modified organism if, say, a toxin gene has been cloned into it. Such an organism would be transferred to Group II under these circumstances. The detailed criteria for classification are listed in the Genetically Modified Organisms (Contained Use) (Amendment) Regulations (1996).

Other organisms. This includes genetically modified organisms other than microorganisms (e.g. animals or plants) and includes transgenic animals. The criteria for the classification of organisms other than microorganisms include:

a) That the organism is not a genetically modified microorganism.
b) That genetically modified non-microorganisms (including whole plants and animals) must be classified into one of two hazard groups – i.e. those that are as safe in the containment facility as any recipient or parent organisms (and as such may be considered as equivalent to a Group I organism), and those that are not. They do not need to be categorized as Type A or B.

Apart from the regulations themselves, a series of guidance notes is also available which includes guidance on work with transgenic animals. These guidance notes, which were originally in separate sections 1–11 (of which Note 9 dealt with transgenic animals), have been reviewed and reissued as a single handbook under the title, *Compendium of Guidance from The Health and Safety Commission's Advisory Committee on Genetic Modification*. The compendium needs to be read in conjunction with other more recent guidance relating to animal units. At the time of publication the revised unified handbook is not fully complete with respect to risk assessment of work with genetically modified animals (HSE/Smith personal communication) and it is strongly suggested that animal users keep in touch with their subcommittee for genetic modification for current information.

Notification of the intent to use premises

The production and use of transgenic animals for research or teaching purposes would normally classify as an operation under the group 'Other organisms'. As such, the intention to use premises for the first time would require notification to the HSE 90 days in advance of commencing work. The use of the premises would normally thereafter begin, unless the HSE objects in writing giving reasons before the end of the 90-day period.

Notification and recording of individual activities

For 'Other organisms' activities it is sufficient:

a) to keep a copy of the risk assessment;
b) to keep a record of work done;
c) to make a retrospective return to the HSE at the end of each calendar year.

Standards of occupational and environmental safety and containment

As for work with any experimental animals, in any activities involving genetically modified animals the principles of good occupational safety should apply, the latter being covered by the Health and Safety at Work etc. Act (1974) and subsequent legislation (see relevant chapters). Particular attention should be given to the Control of Substances Hazardous to Health Regulations (COSHH) (1999), Approved Codes of Practice (1999), which implement the Biological Hazards Directive 90/679/EEC as amended. The regulations specify the control measures to be applied in all occupations for preventing or minimizing the risk of illness associated with exposure to biological agents. Special attention should be given to Schedule 3 of COSHH (1999) – which defines special conditions relating to biological agents – particularly that section dealing with the assessment of health risk and containment measures.

Formation of a local genetic modification safety committee

Before notifying the HSE of the intention to begin work for the first time, it will be necessary for the organization to form a local genetic modification safety committee. Although this is a regulatory requirement, the composition of the committee is not. Members of the committee should be representative of the staff involved and should have suitable expertise to be able to advise on the biological safety of the work. Co-opted members might be appropriate to assess particular proposals or supplement internal expertise.

A necessary function of such a committee before notification of the commencement of work, would be to consider and advise on a risk assessment of the work envisaged and to advise on the proposed level of containment required. This risk assessment could relate to the animal strain itself, but could also include any vector or system used in the generation of the animal. The level of containment will thus depend on:

a) the possibility of the transmission of the gene to related or other species;
b) the possibility of transmission or activation of viruses or other vectors;
c) the possibility of operator exposure to the gene product or any viral vector;
d) possible environmental impact.

Animal containment levels

These are set out under part 3 of the compendium mentioned earlier in the chapter. In outline, where it is proposed to inoculate animals with viable genetically modified microorganisms, animal containment corresponding to that used in the laboratory for the microorganisms concerned should be used (see part 3C of the compendium). Further details of appropriate animal containment levels are also specified in the Advisory Committee on Dangerous Pathogens publication *The categorisation of biological agents according to hazard and categories of containment* (ACDP 1995). This edition reiterates those parts of the COSHH Regulations (1994) concerning containment measures to be used in laboratories and animal rooms and in the industrial use of biological agents (latest COSHH 1999). In general, an animal house which satisfies the requirements of the Animals (Scientific Procedures) Act (1986) and its associated codes of practice would satisfy the recommendations for containment level 1. Where specific animal pathogens are being used in conjunction with transgenic animals, advice should also be sought from the Divisional Veterinary Officer (DVO), since Ministry of Agriculture, Fisheries and Food (MAFF) requirements could also influence the level of containment.

Usually transgenic animals produced by micro-injection or by the use of a replication defective retrovirus vector or other sequences not horizontally transmitted would be allocated to level 1 containment. Those transgenic animals generated by using other viral vectors capable of being shed would normally be contained at a minimum level 2. However, although this is the theoretical approach based on safety, wherever possible consideration should also be given to keeping transgenic animals in a manner designed to protect valuable strains from secondary infection. This could include the use of filter-top cages, filter racks or flexible film isolators. Consideration also needs to be given to the uncertain health status of transgenic animals from many sources, which may themselves prove a hazard for other colonies already present, or the staff using them.

Risk assessment should take into account the recommendations of the relevant guidance notes concerning the use of viral vectors and other modified microorganisms or potentially dangerous substances (see also the Genetically Modified Organisms (Risk Assessment) (Records And Exemptions) Regulations (1996).

Transgenic farm animals

The Advisory Committee on Novel Food Processes (ACNFP) should be notified of proposals to work on transgenic food production animals at an early stage. Some transgenic farm animals such as sheep and pigs are used for the production of pharmaceuticals or for organ transplants. However, the ratio of experimental failure is high and there may be a wish to pass such experimental failures into the food chain.

Where there is a possibility that transgenic animals might be used for human food, it will be necessary initially to inform the ACNFP giving details of the nature of the insert, the vector system, methods for detecting the insert and its activity, the nature of the product itself and any potential risks to humans after consumption. If the ACNFP advises that there is no risk, then normal salvage by sale and slaughter would be possible. If not, or if the producer does not wish to follow this route, then such animals should be killed and incinerated. The disposal of experimental animals for human use is also dependent on the agreement of the Home Office inspector and conditions will normally be set out in the project licence.

The role of the local genetic modification safety committee

The constitution of the local committee has already been outlined, together with its function of advising on risk assessments on the work to be done. However, its relationships with the main safety committee will depend on local circumstances. Some factors do need further explanation.

1) If a supervisory medical officer is a member of the committee, he/she does not necessarily need to attend all meetings, but could simply receive agendas and minutes. Attendance might well be on an 'as necessary' basis.
2) It may be thought useful to have a member from another department to provide an outside viewpoint, particularly where genetic modification might be looked at with some suspicion by other departments resident in the same building.
3) It is useful to review the membership of the committee annually.
4) The committee should meet sufficiently frequently to be able to review all proposals for new work or to reassess ongoing work.
5) The committee should be able to take an overview of general laboratory practices and safety procedures, including accidents and incidents, and to advise the head of department on the needs for training or contingency plans to cope with accidents or accidental release or escape of transgenic animals.
6) The committee is responsible for drawing up local rules and standard procedures for the conduct of the work .
7) The agenda and minutes of the committee should not be considered as confidential documents and should be available for scrutiny as required. It should also be possible for any person having a reasonable personal concern to attend the meetings of the committee.

8) The committee should keep up-to-date on the current legislation and advisory notes and maintain good liaison with the main safety committee and the Health and Safety Executive.

Import and export of transgenic animals

Import

Any importer of transgenic animals, or for that matter anyone bringing in transgenic animals from external sources, should ensure before delivery that adequate facilities are available for the receipt and accommodation of the animals and that the containment levels reflect this. The production and use of transgenic animals is also subject to the requirements of the Animals (Scientific Procedures) Act (1986) administered by the Home Office. Under these circumstances, animals being obtained from outside the UK will be coming from sources not covered by a Home Office Certificate of Designation. As such, special Home Office permission will be required and a Home Office transfer certificate will need to be completed.

Similarly, because the production and use of transgenic animals will involve the use of procedures within the meaning of the Act, it will be necessary for those acquiring them to have the necessary personal and project licence cover.

If the animals are to be obtained from sources outside the European Community, it will also be necessary for them to undergo the statutory six months import quarantine on arrival under the terms of the Rabies (Importation of Dogs, Cats and Other Mammals) Order (1974). In the case of rodents, the time period can to some extent be shortened, by agreement with the Divisional Veterinary Officer, by importing breeding nuclei and removing the offspring of healthy imported animals 15 days after weaning, to a level of containment appropriate for the animal species and strain.

Import from EC countries comes under the control of the Balai Directive (92/65 EEC) implemented in the UK by the Animals and Animal Products (Import and Export) Regulations (1995).

As such it is unnecessary for imported research animals to undergo statutory quarantine providing they originate from premises registered under the terms of 92/65 EEC and have been born and reared there since birth. Nevertheless, they will still be subject to Home Office requirements for animals originating from premises outside the UK which do not come from premises covered by a Certificate of Designation under the Animals (Scientific Procedures) Act (1986). It is also likely that the supplier in the EC country of origin will expect the recipient to sign an undertaking that the transgenic animals are going to persons and premises which comply with the UK Genetically Modified Organisms (Contained use) Regulations already mentioned.

Export

In non-EC countries regulations will vary and it will be a matter for the exporter to establish from the relevant ministry of the country to which the animals are being exported, what that country's requirements are. Exports to EC countries are subject to the same EC Directive 92/65 as for imports. However, under any circumstances, the export of laboratory animals of any kind effectively removes them from the control of the Home Office and will require official permission.

Health surveillance of staff

There is no specific requirement for health surveillance under legislation covering genetic modification. However, other regulations such as COSHH will require a review of the health of those working with genetically modified animals as part of the supervision maintained for animal-related work in general.

Accidental release of transgenic animals to the environment

Special notification and permission is required if a transgenic animal is to be intentionally introduced to the environment (Genetically Modified Organisms (Deliberate Release) Regulations 1995). It should also be noted that the release of any animal not normally resident in the wild state requires the issue of a licence under the Wild Life and Countryside Act (1981). A true escape of a transgenic animal or animals to the environment from a contained unit should be reported to the local genetic modification safety committee, the local Safety Officer and the HSE as soon as possible after the event. It is, however, accepted that the majority of transgenic animals are purpose-bred rodents used only in a controlled environment and for this reason are highly unlikely to escape or to survive in the wild. They are even less likely to breed, particularly since many strains are more susceptible to stress and secondary infection and are thus highly unsuited to life in the wild.

Where animals are kept in such circumstances that escape is possible or likely, e.g. fish farmed in conventional units, the user will be expected to obtain an official 'Consent to release' authority, as the likelihood of escape is high. Where transgenic farm animals are concerned, a farm animal confined in a field protected by stock-proof fencing is not considered as having been intentionally released to the environment. However, it is expected that it or they will be properly supervised, so that any escape is promptly detected.

Marking

All transgenic animals and experimental failures (i.e. those animals in which the gene transfer has not been successful or has not produced the desired result) should be clearly and easily identifiable.

The DNA itself may be used as a marker where no other convenient means of identification exists. Once the animal is no longer in the experimental stage, e.g. it has been released from Home Office control or accepted as an animal for food production, no further special identification is required.

Records

Full records of the preparation, breeding, movement, release and/or disposal of all transgenic animals should be kept. Similarly, under the Animals (Scientific Procedures) Act (1986) the production and breeding of transgenic animals are classified as procedures for which project and personal licences are required. Records will therefore also be required for Home Office purposes.

References

Advisory Committee on Dangerous Pathogens (1995) *The Categorisation of Biological Agents according to Hazard and Categories of Containment*. 4th edn. Sudbury: HSE Books

Animals and Animal Products (Import and Export) Regulations (1995) Statutory Instruments No. 2428. Amended 1996 (SI. 1111). London: HMSO

Animals (Scientific Procedures) Act (1986) London: HMSO

Breitman ML, Clapoff S, Rossant J, Tsui LC, Glade LM, Maxwell IH, Bernstein A (1987) Genetic ablation: Targeted expression of a toxic gene causes microphthalmia in transgenic mice. *Science* **238**, 1563–65

Burke DT, Carle GF, Olden MV (1987) Cloning of large segments of exogenous DNA into yeast by means of artificial chromosome vectors. *Science* **236**, 806–12

Control of Substances Hazardous to Health Regulations (COSHH) (1999) Approved Codes of Practice (1999) L5. Sudbury: HSE Books

Chen TT, Powers DA (1990) Transgenic fish. *Trends in Biotechnology* **8**, 209–15

EC Directive 90/219/EEC. *On the contained use of genetically modified micro-organisms*

EC Directive 90/220/EEC. *On the deliberate release into the environment of genetically modified micro-organisms*

EC Directive 90/679/EEC. *On the protection of workers from risks related to biological agents at work*

EC Directive 92/65/EEC (the 'Balai' Directive) *On the importation of animals and animal products*

Environmental Protection Act (1990) London: HMSO

European Communities Act (1972) London: HMSO

Evans MJ, Kaufman M (1981) Establishment in culture of pluripotent cells from mouse embryos. *Nature* **292**, 154–6

Genetically Modified Organisms (Contained Use) Regulations (1992) Statutory Instruments No. 3217 as amended 1996 (967), 1998 (1548) and Associated Guidance Notes. London: The Stationery Office

Genetically Modified Organisms (Deliberate Release) Regulations (1995) Statutory Instruments No. 304. London: HMSO

Genetically Modified Organisms (Risk Assessment) (Records and Exemptions) Regulations (1996) Statutory Instruments No. 1106. London: HSE

Gordon JW, Ruddle FH (1981) Integration and stable germline transmission of genes into mouse pronuclei. *Science* **214**, 1244–6

Health and Safety at Work etc. Act (1974) and associated regulations. London: HMSO

Hammer RE, Pursel VG, Rexroad CE, Wall RJ (1985) Production of transgenic rabbits, sheep and pigs by microinjection. *Nature* **315**, 680–3

Hogan B, Costatini F, Lacy E (1986) *The manipulation of the mouse embryo.* New York: Coldspring Harbor Laboratory

Jaenish R (1976) Germline integration and Mendelian transmission of the exogenous Molony leukaemia virus. *Proceedings of the National Academy of Sciences* **73**, 1260–4

Pachnis V, Pevny L, Rothstein R, Constantini F (1990) Transfer of a yeast artificial chromosome carrying human DNA from *Saccharomyces cerevisae* into mammalian cells. *Proceedings of the National Academy of Sciences* **87**, 5109–13

Rabies (Importation of Dogs, Cats and Other Mammals) Order (1974) Statutory Instruments No. 2211 as amended. London: HMSO

Thomas KR, Folger KR, Capecchi MR (1986) High frequency targeting of genes to specific sites in the mammalian genome. *Cell* **44**, 419–28

Valerio D (1992) Retrovirus vectors for gene therapy procedures. In: *Transgenic Animals* (Grosveld E, Kollias G, eds). London /New York: Academic Press, pp 211–39

Wildlife and Countryside Act (1981) London: HMSO

5

Chemical hazards

Ian Palotai

Contents

Introduction

This chapter describes the general nature of chemical hazards in the animal house and methods for controlling these hazards. As the presence of specific chemicals will vary from unit to unit, no attempt will be made to deal with individual chemicals in detail. Such health and safety information is available from manufacturers. However, widely applicable principles of chemical hazard control will be considered and illustrated, with examples where appropriate.

A variety of substances hazardous to health may be used in the animal house, ranging from substances used for cleaning and disinfection with well known properties, to metabolites of new pharmaceutical products under test whose properties are unknown. Hazards and risks arise from the work activities of several groups of staff. Prevention of exposure of staff to chemical hazards requires the cooperation of all those using the animal house, under the overall control of the animal house manager who must maintain a high standard of working practices and compliance.

In the United Kingdom the use of hazardous chemicals at work is controlled by a broad range of legislation. The most important requirements are detailed in the Control of Substances Hazardous to Health Regulations 1999 (COSHH) (Statutory Instruments 1999, No. 437) and supporting Approved Codes of Practice (Health and Safety Commission 1999). These require all work with substances hazardous to health to be assessed and adequate precautions taken to control risks to health. Substances hazardous to health are defined in the regulations and, as well as manufactured chemicals, include biological agents that may cause infection, allergy, toxicity or otherwise create a hazard to human health (see other chapters in this handbook).

In all areas of health and safety there is increasing agreement on how hazard and risk are related. For chemicals, the following definitions are illustrative (Council Directives 1994):

- *Hazard* – the intrinsic property of a chemical agent with potential to cause harm. The hazard depends only on the nature of the chemical.
- *Risk* – the likelihood that the potential for harm will be attained under the conditions of use and/or exposure.

Categories of chemical hazard

The classification of chemical substances and preparations according to hazard is a requirement of the Chemicals (Hazard Information and Packaging for Supply) Regulations 1994 (CHIP) (Statutory Instruments 1994, No. 3247 and amendments). The hazardous properties of chemicals must be evaluated by suppliers and information given to users in the form of datasheets and adequate labelling which includes an indication of danger. In the case of new chemical entities, properties must be deduced as far as possible from those of related compounds until further information is available. The categories of hazard include those shown in Table 1.

Chemical hazards associated with laboratory animal care and use

The main natural chemical hazards associated with animal work are allergens in animal products such as urine, serum, dander and saliva. These are covered in Chapter 2 of this handbook. Sterilants, disinfectants, detergents, corrosive cleaning chemicals, veterinary drugs and preparations and chemicals on test in experiments are also encountered. Exposure to hazardous chemicals that have been applied to animals may arise from contact with metabolites in animal excretions and secretions, including through the skin and in exhaled gases, as well as during the initial application which may contaminate the work environment.

Chemical risk assessment

In the UK the COSHH Regulations detail minimum legal requirements for work with hazardous substances and define hazardous chemical substances with reference to the CHIP classifications. The key regulations concern risk assessment and control of exposure and the following list indicates the steps to be undertaken to comply with COSHH (Health and Safety Executive 1993):

Regulation 6
Assessment of health risks
Regulation 7 Prevention or control of exposure
Regulation 8 Use of control measures
Regulation 9 Maintenance, examination and testing of control measures
Regulation 10 Monitoring exposure
Regulation 11 Health surveillance
Regulation 12 Information, instruction and training

In assessing the risks from using substances hazardous to health, the most successful approach is to look systematically at the tasks that are undertaken in the animal house. It should be a supervisory

Table 1 Classification of chemical hazards

Category of danger	Property	Symbol letter	Symbol
Very toxic	Substances and preparations which in very low quantities cause death or acute or chronic damage to health when inhaled, swallowed or absorbed via the skin.	T+	
Toxic	Substances and preparations which in low quantities cause death or acute or chronic damage to health when inhaled, swallowed or absorbed via the skin.	T	
Harmful	Substances and preparations which may cause death or acute or chronic damage to health when inhaled, swallowed or absorbed via the skin.	Xn	
Corrosive	Substances and preparations which may, on contact with living tissues, destroy them.	C	
Irritant	Non-corrosive substances and preparations which, through immediate, prolonged or repeated contact with the skin or mucous membrane, may cause inflammation.	Xi	
Sensitizing	Substances and preparations which, if they are inhaled or if they penetrate the skin, are capable of eliciting a reaction by hypersensitization such that on further exposure to the substance or preparation characteristic adverse effects are produced.	Xn or Xi	
Carcinogenic	Substances and preparations which, if they are inhaled or ingested or if they penetrate the skin, may induce cancer or increase its incidence.	T or Xn	
Mutagenic	Substances and preparations which, if they are inhaled or ingested or if they penetrate the skin, may induce heritable genetic defects or increase their incidence.	T or Xn	
Toxic for reproduction	Substances and preparations which, if they are inhaled or ingested or if they penetrate the skin, may produce or increase the incidence of non-heritable adverse effects in the progeny and /or an impairment of male or female reproductive functions or capacity.	T	

duty to examine critically each procedure to decide where hazardous substances are involved, to identify whether exposure to the substances is likely and to assess the potential consequence of such exposure. It is essential that this process should involve staff who have practical experience of the work in question and assessment should include the following steps:

- Obtaining information on the substance.
- Identifying how exposure to the chemical could occur during the work and how severe any exposure is likely to be.
- Deciding whether working practices are adequate to control the risk of exposure.
- Changing working practices where necessary to ensure that the risks to health are eliminated or reduced and that staff know and understand safety precautions.

Hazard information on chemicals used at work is obtainable from the supplier of the chemical, who has a duty to provide safety information under Section 6 of the Health and Safety at Work etc. Act 1974. The exact type of information required is set down in the CHIP regulations in 16 categories. For proprietary products such as cleaners and disinfectants containing substances hazardous to health, information should relate to the properties of the formulation and not to the pure chemical ingredients.

Information on the effect of hazardous chemicals on the body is necessary in order that staff can be instructed to recognize symptoms of exposure. This is important in case engineering, e.g. ventilation, or other control measures fail or provide only partial control of exposure. This may, for example, occur during gaseous anaesthesia of non-intubated animals. Staff should understand what remedial steps are to be taken in these circumstances. Particular care is necessary where substances have either little odour or an odour that dulls the sense of smell on prolonged exposure, as with certain anaesthetics. When staff are working alone with these substances, checks by others may be appropriate from time to time to ensure workroom levels of anaesthetic have not risen unexpectedly.

For many industrial chemicals, exposure limits are set. These have been published in Schedule 1 to the COSHH regulations and in the annual Guidance Note EH40 (e.g. Health and Safety Executive 1999), but henceforth will be given in EH40 only (see Statutory Instruments 1999, No. 437). Guidelines for setting exposure limits for pharmaceuticals are also available (The Association of the British Pharmaceutical Industry 1992, 1995). The criteria used for setting these limits include, where available, an evaluation of data from animal toxicity studies and refer to exposure to pure chemicals by inhalation. Exposure limits are given in terms of Occupational Exposure Standards (OESs) and Maximum Exposure Limits (MELs). These cover reference periods of 15 minutes or 8 hours and are given in parts per million (ppm) and milligrams per cubic metre (mg m^{-3}). For substances with an OES, a level is set where there is no indication of risk to health. This is based on current scientific knowledge which can change with time. Substances assigned a MEL either do not have a 'safe' exposure limit determined, or have safe levels but control to these levels is not reasonably practicable. These limits are set by the Health and Safety Commission.

With reference to animal allergens, setting exposure limits is not yet feasible. See Chapter 2 for more information on current tests to measure allergens in the environment. The most harmful route of exposure to allergens is thought to be by inhalation of coarse dusts contaminated with animal products such as urine. Measurements of dust levels in the workplace can be carried out and are useful in identifying tasks that generate most dust in the air. It is, however, extremely difficult to relate levels of dust in the breathing zone of staff to the degree of hazard. In this case a reasonable approach by the animal house manager is to direct all staff working with animals such as rats, guineapigs and mice to wear respiratory protection to prevent or at least reduce exposure of the respiratory tract to allergen-contaminated dust.

As well as chemical risk assessment, consideration should be given to other hazards such as radiochemical hazards, biohazards, physical hazards, or flammability or explosion hazards. All should be taken into account when devising control measures for a particular piece of work or activity.

Unless chemical risk assessments are very simple and easily repeated, they should be written down. The assessment process should allow decisions to be made concerning the adequacy of control measures. The written assessment will show that this process has taken place and should detail the precautionary measures necessary to comply with the regulations. The assessment can be incorporated into experimental protocols or operating procedures within the animal unit.

Assessments of work activities involving chemicals should be reviewed at least every five years. In a research environment where work activities change frequently, related risk assessments must be reviewed to take account of such changes. Review should also take place if exposure limits are changed.

Routes and mechanisms of exposure

In assessing the safety of work activities, the routes by which chemicals can gain entry to the body are important. The route influences both engineering control measures and selection of personal protective equipment to prevent or minimize exposure. Planning work activities to avoid the spread of the chemical into the working environment or beyond is also necessary.

Workplace exposures to chemicals in liquid form are either through contact with the skin or eyes, or by inhalation of aerosols or vapours. Ingestion is unlikely to occur accidentally except when splashes involve the mouth or contamination is via hands or gloves.

In many tasks in the animal unit, skin contact with chemicals is the most likely route of exposure. The permeability of the skin to the chemical and the extent of contamination are important factors in determining the seriousness of skin exposure. Eyes are particularly vulnerable to chemical splashes.

Skin exposure may result in localized skin damage or uptake of the chemical through the skin followed by injury to deeper tissues or a different target organ. Many chemicals remove the skin's oily protective coating, increasing skin permeability and increasing the severity of chemical exposure. Repeated exposures may affect the functioning of the skin for as long as the exposure continues giving rise to dermatitis. Causes of dermatitis in veterinary practice have been examined (Falk *et al.* 1985). Exposure to certain antibiotics and detergents are responsible for some cases. Less frequently, repeated exposures may bring about skin changes that result in allergic responses to further exposures.

Exposure of the respiratory tract to airborne chemicals is an important route of exposure. Aerosols are suspensions of very fine droplets of liquid (or particles) in the air which, unlike larger droplets, do not settle quickly onto surfaces. Instead the droplets receive sufficient energy from the surrounding air to be carried in the air until they collide with an object.

Aerosols are produced when liquids are forced through fine openings or strike hard surfaces or are subjected to vibration. Sources of aerosols in the animal house include operations involving water hosing of surfaces when chemical and biological materials may become aerosols. Any procedure involving filling syringes and injection of liquids into animals may give rise to aerosol during needle withdrawal. Although this is unlikely to give rise to airborne aerosol, it is possible that areas surrounding the site of injection will be contaminated with chemical. Homogenization of tissues may also be a source of aerosol contamination.

Dusts may give rise to chemical exposure either following settling on the skin or following entry to the respiratory tract. Large dust particles settling on the skin may be absorbed, particularly when trapped against the skin inside ill-fitting protective clothing such as gloves or sleeves. Large dust particles that are inhaled are captured in the upper respiratory tract and, although unlikely to reach the lungs, may be swallowed eventually and deposited in the digestive system. Smaller dust particles can reach the lungs where local damage or absorption can occur.

The route by which gases enter the body is by inhalation. Gases such as chlorine and sulphur dioxide cause local irritation and damage, whereas gases such as nitrous oxide exert their effect after absorption into the bloodstream by action on the central nervous system. Asphyxiant gases such as nitrogen and carbon dioxide interfere with gaseous exchange in the lungs, resulting in reduced oxygen transfer by the blood through the body. Where human exposure to low concentrations of anaesthetic gases over a period of hours is possible, risk assessments should be made with special care. Typical workroom concentrations of gases should be established and related to exposure limits.

The use of sharps is another source of exposure. Injection of small animals usually involves the risk of needlestick injury. Precautionary measures rely on animal-handing skills and anticipation of animal behaviour in addition to skill with equipment. Exposure in these situations might be to strong irritants such as adjuvants, anaesthetics such as urethane, or to biological materials including genetically modified organisms (GMOs). The consequences of self-injection in these procedures should be assessed during planning of the work.

These routes of entry should be considered when planning work. If the work activity provides an opportunity for exposure then the potential severity of that exposure must be evaluated. The severity will depend on the properties of the material, its concentration and formulation, the duration of exposure, and the extent of the exposure.

When especially hazardous materials are used in work activities where the potential for exposure is great, a well-documented procedure should be written and implemented. Examples in the animal unit include all fumigation activities using formaldehyde and sterilization using ethylene oxide.

Health effects

It is common to consider chemicals in terms of their acute and chronic effects on the body.

Acute effects are seen after a single exposure to the chemical. The severity of the effect is usually in proportion to the amount of chemical entering the body, and the delay between exposure and the onset of symptoms of exposure is usually short. Acute effects are often reversible provided that the exposure is not too extreme. An example would be short-term exposure to chlorine gas following accidental mixing of bleach and acid during lavatory cleaning.

Chronic effects result from repeated exposures to smaller amounts of chemical. There may be no acute symptoms associated with this type of exposure and therefore no warning. Chronic effects appear some time after initial exposure to the chemical and are often irreversible. With both chronic and acute effects there is considered to be a relationship between the degree of exposure and the effect. There is also thought to be a threshold of exposure below which no effect is seen.

Certain chemicals are known to be sensitizers. Sensitization is a chronic health problem that affects only some individuals in an exposed population following repeated exposures. Sensitizing chemicals also often exhibit an acute irritant effect. In a sensitized individual further exposure to the sensitizing chemical, even at much lower levels than were necessary to cause sensitization, can cause adverse reactions. The severity of response seems unrelated to the degree of exposure in sensitized individuals and the sensitization is a permanent deleterious health effect.

Certain animal-derived materials contain sensitizing chemicals. Examples include urine and dander. In the animal house, without proper precautions, some staff will succumb to developing a sensitization to these materials. This occupational illness is known as 'laboratory animal allergy'. It is not possible currently to predict which individuals will be susceptible, and the actual or potential severity of the condition once developed may limit the ability of staff to continue working with the sensitizing species of animal. For this reason, systems of work for all staff working with sensitizing species must ensure that exposure to sensitizing animal products is limited (see Chapter 2).

Carcinogenic materials present special difficulties when deciding on appropriate systems of work. Exposure to carcinogens may cause cancer. The transformation of normal cells into cancer cells is a complex change. Unlike exposure to most other substances, there is no safe level of exposure to carcinogens due to the uncertainty concerning any threshold for biological effect. Standards of working practice for work with carcinogens are detailed in a Carcinogens Approved Code of Practice that supports the COSHH Regulations (Health and Safety Commission 1999, ACOP, L5).

A final class of substances to consider is that of chemicals with potent pharmacological activity. These chemicals are encountered in the pharmaceutical industry and in associated animal studies and have specific actions on biochemical processes. Changes in concentrations of natural chemical messengers normally control these processes, but exposure to pharmacologically active chemicals may interfere strongly and can result in physiological changes. In some cases this effect can be seen not only at a cellular level but also in the functioning of the body as a whole. In other cases the effect of exposure may be to alter biochemical processes without any observable change to overall body function. In these situations an evaluation needs to be made of the significance of these changes. Increasingly, exposure limits are being set with these considerations in mind.

Control measures

Where possible, prevention of contamination of the workplace and the staff therein should be normal practice by the use of a fume cupboard, local extract ventilation or other containment measure. Use of personal protective equipment will be the final resort, though simple equipment such as suitable gloves (i.e. impermeable), eye protection and laboratory overalls are likely to be everyday wear.

The COSHH regulations 1999 specify a series of control measures to be used where exposure needs to be reduced or eliminated. The first option to consider is whether the hazardous substance needs to be used at all. If a job can be done in some other way, this should be considered. If this is not possible, then the least hazardous substance should be used that allows the work to be done. An example of this is the replacement where possible of glutaraldehyde in veterinary areas with other, less toxic, disinfectants. The physical form of the material is also a consideration. Where a chemical can be obtained as a powder, or granules or paste, a decision should be made about which formulation presents the least opportunity for exposure during the work procedure. This choice will be influenced by the equipment available in the workplace to handle different types of material. Dilution of chemicals should normally be carried out by adding the chemical to the volume of water required and not *vice versa*. If dilution of concentrated chemicals can be avoided by the purchase of ready-diluted solutions, then risks from chemical exposure are reduced immediately. Automatic dosing of cagewasher chemicals or of hose points with disinfectant from large reservoirs in plant areas will also cut down handling risks.

The way in which the chemical is used should then be considered. When handling solutions of chemicals, siphoning should be considered in preference to pouring. Where surfaces have to be treated with chemical solutions, spraying should be avoided if painting or wiping are acceptable alternatives.

Use of anaesthetic equipment provides one of the more serious potential exposures to chemicals in the animal house. In the UK, the Health and Safety Executive has published guidelines for healthcare workers involved in work with anaesthetics (Health and Safety Executive 1995). Much of the information is useful to making risk assessments of veterinary procedures. Veterinary exposures have also been assessed (Gardner *et al.* 1991).

Where halocarbon reservoirs can be fitted with filling devices that remove the need for pouring, these devices should be fitted because the filling of reservoirs is an operation that gives rise to a high concentration of anaesthetic for a short period. Secondly, if animals can be intubated during anaesthesia then this should be done because intubation reduces the delivery of anaesthetic into the workroom. Finally, active scavenging of the anaesthetic gases should be provided to remove the gases from the workroom. A further source of exposure is during recovery from anaesthesia, particularly where large numbers of animals are involved and the working period is prolonged. Recovery areas should therefore be provided with sufficient ventilation to prevent the build-up of gases.

Anaesthetic scavenging systems limit the exposure of workers to anaesthetic gases and, as these are engineering controls in terms of the COSHH regulations, a performance specification should be defined. A method of checking that this level of performance is maintained over time must be put in place. This involves measurements of extract flow rate in the system. These checks must be carried out at least once in every period of 14 months and the results of the inspections recorded and kept (Health and Safety Executive 1998). Of equal importance in ensuring control of exposure is the correct use of scavenging systems. Operators should know how the system should be used, should check correct functioning on each occasion of use, and should know what to do should fault conditions be recognized, e.g. by the activation of an alarm. Failure of either the anaesthetic extract system or the room ventilation will increase levels of anaesthetic in the workroom. Disposal of waste anaesthetic absorbed in canisters should be according to manufacturers' instructions.

Work where exposure to formaldehyde can occur needs special care. Formaldehyde has a MEL of 2 ppm. In post-mortem rooms, exposure to formaldehyde should be assessed where formalin solutions are in use. Work stations incorporating extraction can reduce workroom levels of formaldehyde. In addition, simple precautions such as reducing the surface area of trays containing formalin solutions and capping specimen pots except during sample transfer also reduce exposure. The effectiveness of these control measures may be evaluated periodically using diffusive sampling badges or other devices.

When setting up the fumigation of rooms, e.g. with formaldehyde, it is likely that respiratory protection will be needed. Rooms must then be sealed and warning notices posted. Ventilation should be controlled remotely to allow shut-down and purging without re-entering the room until cleared of fumigant, as indicated by a remote monitoring device.

High standards of hygiene and cleanliness in the animal unit are necessary for animal welfare and to prevent the transfer of chemicals from one work area to another. All staff should be made aware of room and equipment cleaning procedures and their role in maintaining high standards. Disinfection procedures should be written into local rules for the animal unit. These should include precautions to be taken when making and using disinfectant solutions. Where water supplies for room cleaning are ready dosed with disinfectant, this should be marked beside hose points and backflow into the mains supply should be prevented by appropriate valves. Selection of disinfectants should include consideration of activity against types of pathogen, conditions of inactivation, compatibility with surfaces and stability of solution (Health and Safety Executive 1989, Health and Safety Commission 1991).

Very simple precautions are important in controlling exposure to hazardous materials. Hand-washing before leaving the animal unit is a necessary routine to reduce the likelihood of transfer of chemical contamination to the mouth or to surfaces elsewhere. A high standard of washing and changing facilities is important in encouraging staff to follow good hygiene rules. Areas should be provided for changing out of outside clothes and for keeping overgarments for use in the animal house. As in any other work environment where hazardous substances are used, eating, drinking, smoking and the application of cosmetics should not be allowed in the animal unit.

Personal protective equipment (PPE) must be suitable for use. It should be constructed to recognized standards, fit well, be compatible with other items of workwear and PPE and provide protection for the user. The Personal Protective Equipment at Work Regulations 1992 set out the requirements for PPE and guidance on selection and standards is provided (Health and Safety Executive 1992). Training in how to put on, use, decontaminate and take off PPE is essential.

Users of powered respirators should be trained in a simple procedure to ensure that the equipment is drawing the right amount of air through the filters. In the case of full-face respirators, a good fit is essential – which will preclude the wearing of beards – and they must be tested in use on a regular basis to ensure proper fit. Visual examination of respirators and the reporting to supervisors of any damage should also be instituted. Records of these examinations are required under regulation 9 of COSHH. Storage for respiratory protective equipment should also be provided, together with facilities for filter changing, testing and cleaning of powered respirators.

Provision of information to staff about work involving hazardous substances is a requirement of the COSHH regulations. Information should be given on the risks involved in the work and the precautions to be used. This is not satisfied solely by providing a datasheet for the materials in use. The information should relate to how the material is used and should be conveyed to staff in a way that is practical and readily understood.

Training to ensure chemicals are used properly is also necessary. This is best incorporated into other aspects of showing staff how to carry out particular tasks and should be specific to the materials in use. Training should include how to store chemicals, e.g. by the use of fire-resistant cupboards or spark-free refrigerators for inflammable materials, drip trays to contain leaks with incompatible chemicals in separate trays, and locked cabinets to hold drugs. Compressed gases should be stored in a well-ventilated area, away from heat sources and the cylinders supported to prevent falling over. Staff should be familiar with action to be taken in case of chemical spillage and direct exposure, e.g. skin contamination. Where there are easily recognized symptoms of over-exposure, staff should be made aware of these and the importance of seeking help immediately.

In considering the choice of control measure, each will have advantages and disadvantages. In making choices and devising systems of work, a balance has to be struck between the reduction in risk that can be achieved against the cost in terms of money, time and effort required to achieve that reduction in risk. This is deciding what is 'reasonably practicable', a key consideration in the UK legislative framework for Health and Safety at Work.

Monitoring

A key factor to success in controlling the use of chemicals in the animal unit is the degree of authority the unit head has over users of the facility, who may not be under direct line management control of the unit head. A second key management interaction in ensuring success of animal unit procedures is the ability of animal unit supervisors to be aware of and, where necessary, stop unauthorized use of chemicals in the unit. Procedures for dealing with such situations should be agreed and should include a mechanism for protocol review involving staff from the animal unit and scientists who use the animals and devise the tests. Where the unit head requires additional safety precautions to be included in protocols, this should be done. Authority should also be given to the unit manager to prohibit work within the unit that does not adequately document and reduce risks to the health and safety of staff in the unit.

The COSHH regulations require two types of monitoring in certain situations. Firstly, workplace monitoring should occur where it is necessary to ensure that control measures are working adequately. Workplace monitoring commonly involves sampling airborne concentrations of chemicals. This may be carried out using grab sampling devices which draw air through chemically impregnated glass detector tubes, or using personal sampling devices worn by staff. Monitoring in this way can provide assurance to staff that exposure is being controlled adequately. Records of workplace monitoring must be kept for five years, or for 40 years in the case of monitoring which is representative of the personal exposure of identifiable members of staff.

The second type of monitoring is health surveillance. Health surveillance does not take the place of personal protective measures, engineering controls or workplace monitoring. It is important in situations where exposure cannot be totally prevented because of the nature of the substance and work activity. The key purpose is to detect as early as possible any adverse health effects. Early detection of disease allows steps to be taken to reverse the condition or prevent it becoming more serious.

A health surveillance programme supported by appropriate human resources policies should be in place in animal facilities. The COSHH regulations require health surveillance to be carried out in specific circumstances. These are where an identifiable health effect related to exposure exists, there is a valid technique for detecting the effect, and exposure levels are such that health may be affected. In the animal unit, laboratory animal allergy is such a condition. Surveillance in this case will involve regular enquiries about symptoms, lung function testing and in some cases biological monitoring in the form of blood testing. The procedures for health surveillance will vary according to the type and severity of the health risk. Procedures for managing cases of occupational ill-health involving line management, occupational health professionals and personnel staff should be in place. Records of health surveillance should be retained for a period of 40 years.

The Employment Medical Advisory Service of the Health and Safety Executive can assist employers in identifying occupational health services in any area of the country.

Regular safety inspections of the animal unit allow a systematic examination of working procedures. These should be conducted by staff from within the animal unit and may involve staff from other work areas. The purpose of the inspections is to ensure that safety standards are being maintained within the unit. This relates to aspects of the facility, the materials in use, the way staff are working and the way equipment is being used, serviced and tested. Chemical use and storage can be made the subject of an inspection. Checklists devised by staff working in the unit can assist those involved in inspections provided that they guide rather than control the conduct of the inspection. Reports from inspections should reinforce good standards of working, as well as highlight the need for corrective actions where inadequate working standards are noted.

Emergencies

Spill control is an aspect of chemical usage that should be included in risk assessments of chemical use in the animal unit. Procedures should be in place to deal with spillages that are foreseeable, taking into account the tasks being undertaken in the unit. The most important decision to be made in the case of spillage is whether unit staff are capable of dealing with the spillage or whether the emergency services are required.

The size of likely spillage should be considered together with the hazardous properties of the chemicals concerned. Material safety data sheets provide information on dealing with spillages. Simple irrigation with tap water or sterile eye-wash saline in some cases may be sufficient to deal with splashes to skin or eyes. The precautions needed to protect staff involved in cleaning up spillages will vary according to the severity of the hazard – from no additional precautions up to evacuation of an area and clean up with full breathing apparatus. In dealing with chemical spillages, it is important to prevent further exposure of staff to the material and to clean up the spillage in the right way. A decision must be made quickly about the extent of any evacuation required. Equipment to allow clean up of chemical spillages should be provided in the animal unit for use in these situations. The solution may lie in the use of water, sand or sawdust, but in some cases kits containing several different and separate absorbent materials specific for acids, bases or organic liquids may be appropriate. Staff should be trained in selecting and using the correct type of material for spillage clean up and should understand the limitations of these materials. Appropriate and safe disposal is crucial. The success of staff in dealing with spillages will depend to a large extent on the preparation and rehearsal of spillage control procedures.

As some chemicals are flammable and combustion of plastic, for instance, may give rise to toxic smoke, appropriate fire extinguishers must be available and training given in when and how to use them. The local emergency services, especially the Fire Service, should be made aware of any chemicals or other agents that might affect how they deal with a problem. A coded notice on the outside of a building may be useful in this respect.

Waste management

Waste generated within the animal unit should be segregated, based upon the disposal routes available to the facility. Special consideration should be given to carcass disposal, where contamination with chemicals or radioactive substances must influence the choice of disposal route and method of intermediate storage. All waste from the facility must be disposed of under a general duty of care in accordance with agreements made between the facility and the Environment Agency or Local Authority.

Formal rules and regulations

Much of this chapter has concerned the COSHH regulations applicable in the UK. In the United States the Federal Hazard Communication Standard (Occupational Safety and Health Administration (OSHA), 29 CFR 1910.1200) and the Occupational Exposures to Hazardous Chemicals in Laboratories Standard (OSHA, 29 CFR 1910.1450) specify how chemicals may be used at work. Together these cover:

- Evaluation of chemical hazards.
- Availability of hazard information.
- Labels and warnings.
- Material Safety Data Sheets.
- Training and information.
- Basic rules and procedures.
- Chemical procurement, distribution and storage.
- Use and maintenance of control measures.
- Employee information and training.
- Approval procedures for certain work.
- Medical examination programme.
- Maintenance of records relating to personal exposure monitoring and medical examination.

Conclusions

The COSHH regulations require assessment of work activities that involve the use of substances hazardous to health. Suppliers of chemicals provide hazard information for the chemicals and preparations they supply under the CHIP regulations. The emphasis in carrying out risk assessments is on considering not only the hazard presented by the chemical, but also the potential for and likelihood of significant exposure during the work activity. Control measures to ensure safe use, storage and disposal of chemicals should be in place in the animal unit and arrangements for provision of information, instruction and training should be made. Health surveillance is required where exposure to animal-derived chemicals occurs and procedures for dealing with chemical spillages should be devised. It is unusual in the animal unit to be carrying out work where risks from chemical exposure are the only risks to be considered. Chemical risk assessment is therefore one component of safety management that is required in the animal unit.

References

Council Directives (1994) Amended proposal for a Council Directive on the protection of the health and safety of workers from the risks related to chemical agents at work. *Official Journal of the European Communities* 94/C 191/104

Falk ES, Hektoen H, Thune PO (1985) Skin and respiratory tract symptoms in veterinary surgeons. *Contact Dermatitis* 12, 274–8

Gardner RJ, Hampton J, Causton JS (1991) Inhalation anaesthetics—exposure and control during veterinary surgery. *Annals of Occupational Hygiene* **35**, 377–88

Health and Safety at Work etc. Act (1974) London: HMSO

Health and Safety Commission (1991) *Safe working and the prevention of infection in clinical laboratories.* Appendix 4, Health Services Advisory Committee. London: HMSO

Health and Safety Commission (1999) *COSHH Regulations 1999 – General COSHH ACOP, Carcinogens ACOP and Biological Agents ACOP.* L5. Sudbury: HSE Books

Health and Safety Executive (1989) *The selection and use of disinfectants.* Cottam AN, *Specialist Inspector Report 17*. London: HMSO

Health and Safety Executive (1992) *Personal Protective Equipment at Work Regulations – Guidance on Regulations*, L25. London: HMSO

Health and Safety Executive (1993) *A step by step guide to COSHH assessment. HS(G) 97*. London: HMSO

Health and Safety Executive (1995) *Anaesthetic Agents: Controlling Exposure under COSHH.* London: HMSO

Health and Safety Executive (1998) *The maintenance, examination and testing of local exhaust ventilation. HS(G) 54.* Sudbury: HSE Books

Health and Safety Executive (1999) *Occupational Exposure Limits. Guidance Note EH40* (published annually). Sudbury: HSE Books

Statutory Instruments (1994) No. 3247 (amended 1999, No. 197) *The Chemicals (Hazard Information and Packaging for Supply) Regulations.* London: HMSO

Statutory Instruments (1999) No. 437. *Control of Substances Hazardous to Health Regulations.* London: The Stationery Office

The Association of the British Pharmaceutical Industry (1992) *Guidelines for the Control of Occupational Exposure to Therapeutic Substances.* London: ABPI

The Association of the British Pharmaceutical Industry (1995) *Guidance on Setting In-House Occupational Exposure Limits for Airborne Therapeutic Substances and their Intermediates.* London: ABPI

Further reading

Ashton I, Gill FS (1992) *Monitoring for Health Hazards at Work.* 2nd edn. Oxford: Blackwell Scientific Publications

Kellard B (1994) *Hazardous Substances: Carcinogens Guide.* 3rd edn. Kingston upon Thames: Croner

Luxon SG (1992) *Hazards in the Chemical Laboratory.* 5th edn. London: Royal Society of Chemistry

Sax I, Lewis RJ (1992) *Dangerous Properties of Industrial Materials.* 8th edn. New York: Van Nostrand Reinhold

6

Radiation safety

D M Taylor

Contents

Introduction

Techniques involving the administration of radioactive materials, or the exposure of animals to ionizing radiations from X-ray, γ-ray or neutron sources are now widely employed in biomedical research. In recent years non-ionizing radiation from ultrasound, microwaves and magnetic fields have found increasing applications in biology and medicine. Frequently, the application of these techniques involves the use of experimental animals; consequently, the potential dangers of working with sources of ionizing or non-ionizing radiation must be understood by laboratory and animal house personnel.

Like electricity and many of the chemical substances commonly used in animal houses, radiations arising from X-ray machines, charged particle generators, radioactive materials or other sources, are a potential hazard to human health. Also, like electricity, the potential hazards from X-ray machines and particle generators go away when the machine is switched off, but, like chemicals, the hazards from radioactive materials cannot be switched off and are always present. However, if the potential dangers are clearly understood and all the appropriate techniques and safety procedures are strictly employed in their manipulation, radiation sources probably represent a very much smaller risk to the individual worker than, for example, many commonly used drugs and other chemicals.

Sadly, the toxicity of many toxic industrial and laboratory chemicals was recognized only after a number of people had died, or suffered serious damage to their health. However, in contrast, the ability of ionizing radiations to cause severe biological damage was recognized very soon after their discovery and well before they came into widespread use in medicine or biomedical research. Consequently it has been possible to propose guidelines, sometimes called 'codes of practice', to permit safe working with ionizing radiation sources and with radioactive materials. Nowadays, in most countries of the world these guidelines have been backed up by statutory legislation to control all types of radiation work. This will be considered later. As a consequence of the early recognition of the potential hazards of ionizing radiation and of the introduction of recommended safety procedures, serious radiation injury of workers has been a very rare occurrence during the past 70 years, despite the many thousands of people throughout the world who have used radioactive materials or radiation sources in their daily work.

In order to appreciate the risks involved in handling any toxic agent, and to understand the principles underlying the recommended safety procedures, it is necessary to have a basic knowledge of the properties of the agent and the nature of the injuries that it may produce. The aim of this chapter is to review the basic properties of ionizing and non-ionizing radiations, the nature of the biological damage that they may induce, and the principles on which the recommended procedures for the safe use of radiation sources and radioactive materials are based. In animal houses the work of the experimental scientist and the laboratory animal scientist meet, and indeed often overlap; thus it is essential that each should have a good understanding of the problems of the other. It is especially important that individual responsibilities should be clearly defined and set out in local rules so that mishaps and misunderstandings do not occur.

Basic aspects of radiation and radioactivity

All matter, living or inanimate, is composed of elements of which about 100 are known to exist; the smallest unit of any element which can exist is called an atom. Atoms are made up of sub-atomic particles of which the three most important are called protons, neutrons and electrons. Protons carry a single positive electrical charge, while neutrons – which have almost the same mass ($\sim 1.7 \times 10^{-27}$ kg) as a proton – are electrically neutral. Electrons, which have a mass of about one eighteen hundredth of a proton, carry a negative electrical charge of equal magnitude to the positive charge on the proton. The atoms of each element possess a characteristic structure which may be described in simple terms as a nucleus made up of protons and neutrons surrounded by layers, or shells, of electrons. The numbers of protons and electrons in an atom of any element are equal, so that it is electrically neutral. For any given element the atoms contain a fixed number of protons and electrons, but the number of neutrons in the nucleus may vary. Atoms containing the same number of protons in their nucleus but different numbers of neutrons are called isotopes. The chemical properties of an atom are determined by the number of electrons it possesses, and thus the number of protons in its nucleus; consequently, the isotopes of any element all share the same chemical properties and differ from each other only by their relative atomic mass.

If some external influence causes an electron to be ejected from an atom, it is left with a net positive charge; this process is called ionization, and the resulting charged atom is known as an ion. If the neutron-to-proton ratio in the nucleus of an element falls outside certain limits the nucleus may be unstable and break up, releasing a large amount of energy in the form of electromagnetic radiation, and, usually, one or more of the component particles of the atom. This process is called radioactive decay.

The commonest types of sub-atomic particle or electromagnetic radiation that arise from radioactive decay or are produced by X-ray or other machines are described below. If these radiations or particles hit the human or other animal body, some or all of their energy may be deposited in the tissues and cause biological damage.

Alpha particles

These are helium ions carrying two positive charges. They are heavy particles with very high energies which have very little ability to penetrate matter; for example an α-particle would be unable to pass through a sheet of thick paper. In the tissues of humans and other animals α-particles have a range of $< 100\,\mu m$; nevertheless, they are capable of producing a great deal of damage because they give up the whole of their vast energy in a very small volume of tissue. For this reason α-particles are described as having a high linear energy transfer, or high LET radiation.

Beta particles

β-particles are very high-speed electrons that are emitted from the atomic shell during radioactive decay. Their energy may vary widely according to the way in which they are produced, and consequently their ability to penetrate matter – including biological tissue – may range from a fraction of a millimetre (mm) to a few centimetres (cm). Most β-particles carry a negative electrical charge, but in some isotopes there are too few neutrons in the nucleus for it to be stable and the electrons emitted during the radioactive decay may have a positive charge; such particles are called *positrons*. When a positron and an electron collide with each other the two are destroyed, or annihilated, with the emission of two gamma rays (see below) each of energy 0.51 MeV.

Neutrons

Neutrons are ejected from the nucleus with high energies during certain types of radioactive decay; they may also be produced in cyclotrons or similar electrical machines. Being uncharged, neutrons are not repelled by charged nuclei but may enter into them to form radioisotopes (also called radionuclides) which decay with the emission of α- or β-particles or γ-rays.

Electromagnetic radiation

In addition to the ejection of sub-atomic particles, radioactive decay is often also associated with the emission of γ-rays or X-rays. These two radiations are examples of electromagnetic radiation, which also includes radiowaves, microwaves, visible, infrared and ultraviolet light and consists of 'packets' of energy, called quanta, which are transmitted in the form of a wave motion. The wavelength of electromagnetic radiation covers a very broad spectrum, ranging from radiowaves with wavelengths of a few kilometres (km) to a few centimetres, to γ-rays whose wavelengths are of the order of a picometre (10^{-12} m). The energy of the quantum, usually called the photon, is inversely proportional to the wavelength; thus the photons of radiowaves carry little energy while γ-ray photons have much energy. All electromagnetic radiation travels through space at the speed of light and both γ- and X-rays may have sufficient energy to pass through several centimetres of lead or many tens of centimetres of concrete. When γ-rays pass through the human body, there is generally relatively little deposition of energy and consequently they are described as low linear energy transfer, or low LET, radiation.

γ-rays and X-rays are in most respects identical, the main difference being that γ-rays are emitted as a result of a change in the atomic nucleus, whereas X-rays arise during radioactive decay, or they may be produced by the application of electrical energy in machines such as those used in hospitals to produce X-rays for diagnostic radiography, or for radiotherapy.

Ultraviolet, or UV, radiation is a form of electromagnetic radiation whose wavelengths start at the lower, violet, end of the spectrum of visible light at about 400 nanometres ($1\,\text{nm} = 10^{-9}$ m) and extend down to about 10 nm. The energy carried by the photons of UV radiation is high enough to cause chemical changes which may lead to biological damage in living systems.

Microwave radiation lies in the wavelength region from about 3 mm to 300 cm. This form of electromagnetic radiation is also capable of causing biological damage, not by direct absorption of energy in specific molecules but by local heating of tissues.

Radiation units

The International System of Units (SI) is now the accepted system used in radiation work in most countries; however, many of the old units are still used, both in legislation and in practice in certain countries, for example in the United States of America. For this reason this section describes both the SI units and their old equivalents. However, it should be remembered that SI units are now the preferred ones, and are those used in the United Kingdom and European Union legislation and regulations concerning radiation work.

Radiation energy is usually expressed in terms of the electron volt (eV), which may be defined as the energy gained by an electron when it is accelerated through a potential difference of 1 volt. The eV is a very small unit and the most commonly used units are one thousand electron volts and one million electron volts, which are abbreviated to keV and MeV, respectively.

The SI basic unit of radioactivity is the becquerel (Bq), which is defined as the amount of a radioactive substance which decays at a rate of one disintegration per second. Again this is a rather small unit and for research and medical applications the larger units of the kilobecquerel (kBq), megabecquerel (MBq) and gigabecquerel (GBq), representing one thousand (10^3), one million (10^6) and one thousand million (10^9) Bq, respectively, are used. The former unit was the Curie (Ci), named after Professor Pierre Curie, which represents the amount of a radioactive substance that has an activity of 3.7×10^{10} disintegrations per second; this is a very large unit and it is commonly subdivided into sub-units, the milli- (mCi), micro- (μCi), and nanocurie (nCi), representing, respectively, one thousandth (10^{-3}), one millionth (10^{-6}) and one thousand millionth (10^{-9}) of a Ci. Thus 1 Ci is equivalent to 37 GBq and 1 μCi corresponds to 37 kBq.

For the purposes of radiation protection the fundamental quantity is that of absorbed dose (D), which represents the amount of radiation energy deposited per unit mass of matter. The SI unit of absorbed dose is called the gray (Gy), which is defined as the absorption of 1 joule of radiation energy per kilogram of matter ($1\,J\,kg^{-1}$). The former unit was the rad, which was defined as the absorption of 100 ergs/gram ($0.01\,J\,kg^{-1}$) of matter; thus 1 Gy is equal to 100 rads.

The biological effects of radiation vary with the type as well as the dose of radiation to which the organism is exposed. Thus it is necessary for radiation protection purposes to have a unit of dose which takes account of the type and energy of the radiation causing the dose. This need has led to the concept of the equivalent dose which is defined as the absorbed dose multiplied by a radiation weighting factor, w_R. Thus the equivalent dose, H_T, in a tissue T is given by the expression:

$$H_T = \sum_R w_R . D_{RT}$$

where D_{RT} is the absorbed dose averaged over the organ or tissue T, due to radiation R. The unit of equivalent dose is expressed as J kg^{-1} and has the special name sievert (Sv). For radiation protection purposes the Gy and the Sv are often expressed in the sub-units of micrograys/microsieverts (μGy/μSv) or milligrays/millisieverts (mGy/mSv), representing, respectively, one millionth and one thousandth of a Gy/Sv. Very large radiation doses are often expressed in units of kilograys/kilosieverts (kGy/kSv), or megagrays/megasieverts (MGy/MSv), representing, respectively one thousand and one million Gy/Sv. The values of the radiation weighting factor w_R for the various types of radiation, as defined by the International Commission on Radiological Protection (ICRP 1991) are listed in Table 1.

The former unit of equivalent dose, or rather dose equivalent as it was then called, was the rem, defined as the absorbed dose in rads multiplied by a quality factor Q, which was similar to the current w_R; thus 1 Sv is equal to 100 rem. Table 2 compares the 'new' and the 'former' radiation units.

There are no special units for non-ionizing electromagnetic radiations, such as UV- or microwaves. All radiations obey the same physical laws and they can be described in terms of their wavelength or frequency. Frequency is the more fundamental unit and is generally used to describe microwave and radiofrequency radiations, while wavelength is mainly used to describe UV and infrared. The strength of the radiation is often expressed in terms of

Table 1 Radiation weighting factors

Type and energy range	w_R
Photons of all energies	1
Electrons of all energies	1
Neutrons of energy:	
$<$ 0.01 MeV	5
0.01–0.1 MeV	10
$>$ 0.1–2 MeV	20
2–20 MeV	10
$>$ 20 MeV	5
Protons, other than recoil photons, of energy $>$ 2 MeV	5
Alpha particles	20

Table taken from ICRP Publication 60 (1991) *Annals of the ICRP* **21** (1–3)

Table 2 The important radiation units

	SI units		Former units	
Quantity	Name	Value	Name	Value
Absorbed dose	Gray	1 J/kg	Rad	100 ergs/g
Equivalent dose	Sievert	1 J/kg	Rem	100 ergs/g
Radioactivity	Becquerel	1 dis/sec	Curie	3.7×10^{10} dis/sec

dis/sec = disintegrations per second
J = joules
100 ergs ≡ 0.01 J

the amount of power delivered to a unit area, normally as watts per square metre ($W\,m^{-2}$). The watt is a relatively large unit, thus, exposures are usually expressed in the smaller units of milliwatts ($mW\,m^{-2}$) or microwatts ($\mu W\,m^{-2}$) per square metre.

For electromagnetic radiations the physical units used to define protection limits depend on the wavelength or frequency and are: current density in ampere per square metre ($A\ m^{-2}$); the specific absorption rate (SAR), which is expressed in units of watts per kilogram ($W\,kg^{-1}$); specific energy absorption (for pulsed radiation) (SA) in units of joules per kilogram ($J\,kg^{-1}$) and power density in units of $W\,m^{-2}$.

Biological effects of radiation

Ionizing radiation

When ionizing radiation passes through living tissues it may damage the cells in its path, the amount of damage being related to the amount of energy absorbed. The exact mechanisms by which the deleterious effects are produced are not yet fully understood, but they most likely arise from the chemical reactions associated with the ionization that occurs along the track of the radiation, which in turn damage the complex molecular systems within the cell, especially the genetic material in the cell nucleus (Arena 1971).

When living tissue is exposed to large doses of radiation, a high proportion of the cells in the irradiated area may be killed. This effect is utilized in radiotherapy – the treatment of cancer with X-rays and other types of radiation. The way in which tissues respond to radiation varies widely depending on the types of cell present, the quality or type of the radiation, the rate at which the irradiation occurs, the dose rate, and the number of irradiations. For example, a rat will die within a few days if it is exposed to a single dose of about 5 Gy of X-rays administered in a few minutes, but it will survive for almost its normal life span if it is irradiated continuously with γ-rays of equivalent energy at 0.5 Gy/day. Although they penetrate only a few micrometres into tissue, α-particles cause very intense, localized areas of ionization, which may cause much more severe damage to cells than would the absorption of an equal amount of β- or γ-radiation within the same group of cells.

Much of our knowledge of the biological effects of radiation comes from studies with experimental animals. However, important direct human information has been derived from studies of patients treated for cancer and certain other diseases with X-radiation or γ-radiation, from lifetime studies of the survivors of the two atomic bombs that were dropped on Hiroshima and Nagasaki in Japan in 1945, and from occupational exposure to radiation such as that sustained by uranium miners and workers with radium and radium-containing luminous paints. The effects of radiation on humans can be divided into two classes: early, or acute, effects, such as nausea, vomiting, and reductions in the numbers of leukocytes, platelets and erythrocytes in the blood, which arise within hours or days of exposure to large doses of radiation (more than 1 Gy); and late effects, such as leukaemia, cancer, cataract, sterility and genetic damage which may appear many months or years after exposure to very much lower doses of radiation.

Exposure of the human body to radiation occurs in two ways:

- by *external* radiation arising from sources outside the body, such as X- and γ-rays, high energy β-particles, neutrons; or the cosmic rays which come from outer space.
- by *internal* radiation from radioactive materials which have entered the body naturally in food or water, or as the result of medical treatment or an accident. Accidental intake of radioactive materials may occur through swallowing (ingestion), through inhalation of gases, dusts or aerosols, or by absorption through the intact skin or through a wound.

The ultimate fate of radionuclides (radioactive isotopes) that enter the body depends on their chemical properties and their interactions with the natural components of the cells and tissues. Some materials may become widely distributed within the body, while others may deposit in specific organs; for example, iodine radionuclides localize in the thyroid gland, while those of calcium, strontium and radium deposit predominantly in the skeleton. The fate of a radionuclide in the body may also be influenced by the chemical form in which it enters the body. For example, carbon-14 administered for medical investigation in the form of ^{14}C-labelled cortisol is very rapidly excreted from the body in virtually unchanged chemical form, whereas ^{14}C-labelled glycine undergoes metabolic conversion in the body to enter fats, proteins and other substances and some fraction of this ^{14}C may be retained for many months, or even years. Radionuclides which deposit in the skeleton may be retained for more than 50 years.

The ability of a radionuclide to cause biological damage after entry into the human body depends not only on its biological behaviour but also on its physical half-life, that is the time required for the radioactivity to decay to one half of its initial value, and the types and energies of its radiations. For example, radio elements which deposit predominantly in bone, and which emit α-particles, such as ^{239}Pu, ^{226}Ra, or ^{228}Th, are regarded as highly radiotoxic, while others, such as ^{3}H (tritium), are regarded as of low toxicity.

It is important to recognize that the human body has been exposed to radiation from natural sources, cosmic rays, γ-rays from rocks and soil, and from α-, β- and γ-radiation from naturally occurring radioactive materials in our food and drink throughout our evolution on this planet. In the United Kingdom this natural radiation amounts on average to 2.20 millisieverts per year (mSv/a), of which the natural radioactive gas radon, and its radioactive daughter products contribute about 60% (Hughes 1993). In addition to this natural irradiation, the UK population is exposed to further irradiation averaging 0.37 mSv/a from medical X-ray exposures and a further 0.1 mSv/a comes from occupational exposure and other sources such as waste discharges and nuclear weapons fall-out. The magnitude of the natural radiation exposure varies in different parts of the world; for example, at high altitudes the cosmic ray component is higher than at sea level, and the contribution from the γ-rays emitted by natural radio elements in rocks, etc., is higher in areas rich in uranium and thorium-containing minerals, such as granite rocks or monazite sands, than in other areas.

Non-ionizing radiations

Non-ionizing radiations with wavelengths of less than a few centimetres have only limited penetration into the human body and their effects are generally limited to surface structures such as skin and the eyes. Penetration increases as wavelengths lengthen, so that energy may be deposited in deeper tissues. The known harmful effects are photochemical with UV radiation, and thermal (heat) with microwave radiation.

Unlike the biological damage produced by ionizing radiations, which results from chemical reactions occurring along the track of the radiation, UV produces chemical effects by direct deposition of the energy of the UV photon in specific molecules. There is a threshold photon energy for each type of reaction, below which it will not occur. In some cases the photochemical reactions initiated by UV radiation are beneficial to humans or animals – for example, the conversion of pro-vitamin D to vitamin D – but most are potentially harmful. Exposure of the unprotected eye to UV radiation can cause severe inflammation, conjunctivitis, associated with intense local pain, and even temporary blindness (snow blindness); however, recovery normally follows within about 24 hours. UV irradiation of the skin may result in acute effects such as erythema or burning (sunburn). Prolonged, or chronic, irradiation of the skin may cause premature ageing of the cells, the formation of pre-cancerous changes or cancer of the skin in the form of basal cell tumours or melanomas. In some individuals exposure to UV radiation may cause hypersensitivity reactions such as the so-called 'photochemical dermatitis' or urticaria.

Exposure to electromagnetic radiation at frequencies greater than about 50 megahertz (MHz) may cause damage due to the direct deposition of heat in the tissue. At microwave frequencies ($\sim$1 to $\sim$300 gigahertz [GHz]) heating results from the alignment and relaxation of molecules, predominantly water molecules, in the tissues causing a feeling of warmth, cell excitation effects and changes in ion fluxes across membranes. Although changes in the rates of cell division and cell transformation are known to occur, there is at present no convincing evidence to suggest that electromagnetic fields cause genetic damage or are able to initiate cancer. High levels of exposure of the eyes to microwave radiation may cause cataract-like changes leading to localized areas of opacity in the lens of the eye; these result from local heating in the tissue which lacks a direct blood supply to carry the heat away. Skin burns may occur after exposure to relatively large doses of microwave radiation. The most sensitive indicator of exposure to microwave radiation in animals is the temporary disturbance of learned behaviour; this effect occurs at high specific absorption rates (SAR), 2–8 watts/kg ($W\,kg^{-1}$).

The philosophy of radiation protection

The potential for X-rays to cause acute adverse effects was recognized within a few months of their discovery in 1895, and by the early 1920s the British X-ray and Radium Committee and the American Roentgen Society had proposed general radiation protection recommendations for avoiding acute effects. These activities led to the formation in 1928 of the International X-ray and Radium Protection Committee, which in 1950 became the International Commission on Radiological Protection or ICRP as it is widely known. The terms of reference of the ICRP are to advance for the public benefit the science of radiological protection and, in particular, to provide recommendations and guidance on all aspects of radiological protection. In preparing its recommendations the ICRP considers the fundamental principles and quantitative base upon which appropriate radiation protection measures can be established. However, it should be emphasized that ICRP makes recommendations only. It is the responsibility of governments and their protection bodies to formulate the specific advice, codes of practice and regulations which are best suited to their national needs.

Over the years the ICRP has revised its recommendations in the light of advancing knowledge concerning the effects of radiation on human beings. The latest Recommendations, which were published in 1991 (ICRP 1991), set out the philosophy underlying the recommended conceptual framework of radiation protection, and also review the scientific evidence upon which this is based. The system of radiological protection recommended by ICRP is based on three general principles:

- **Justification** of a practice, which means that no practice shall be adopted unless it produces sufficient benefit to the exposed individuals or to society to offset the potential damage – the radiation detriment – which it causes.
- **Optimization** of protection by ensuring that within any practice the magnitude of individual radiation doses, the number of people exposed, or potentially exposed, should be kept as low as reasonably achievable, economic and social factors being taken into consideration. This is often called the ALARA Principle, the abbreviation coming from the initial letters of **A**s **L**ow **A**s **R**easonably **A**chievable. To achieve this, exposures should be constrained by restriction on the doses to individuals, or the risks to potentially exposed persons, in order to limit the inequity likely to result from the inherent economic and social judgements.
- **Individual dose and risk limits**. The exposure of individuals resulting from the combination of all the relevant practices should be subject to dose limits, or to some control of risk to potentially exposed persons. These are aimed at ensuring that no individual is exposed to radiation risks that are judged to be unacceptable from these practices in any normal circumstances.

The detrimental effects of radiation against which it is necessary to provide protection are now called deterministic and stochastic effects. Deterministic effects result from the killing of cells which, if the dose is large enough, will be sufficient to impair the function of the tissue; an example is cataract, the opacification of the lens of the eye. The effect is not seen unless the radiation exposure exceeds a certain, threshold level; for protection against such effects the dose limit is set below the threshold. Stochastic effects arise when a cell is modified but not killed, and such modified cells may – after a prolonged delay – develop into, for example, a cancer. Stochastic effects are believed to occur, albeit infrequently, even at the lowest radiation doses and therefore have to be taken into account at all doses. If the modification occurs in a cell whose function is to transmit genetic information to later generations – for example in ova or spermatozoa – any resulting effects, which may be of many different kinds and severity, are expressed in the children of the exposed person. This type of stochastic effect is called an 'hereditary' effect. On the basis of the latest available evidence from all sources, the ICRP has calculated probabilities of stochastic effects arising in exposed individuals and these are shown in Table 3. The Commission has also assessed the distribution of detriment in the different organs and tissues by considering first the fatal cancer probability in each of them, multiplying by an appropriate factor for non-fatal cancer, adding in the probability of severe hereditary disease and adjusting for the relative length of life lost. This distribution of detriment among organs is represented by the tissue weighting factors, w_T, which are listed in Table 4. These values are used for the calculation of the effective dose (sieverts) in the tissue or organ using the expression:

$$\mathrm{E} = \sum_{\mathrm{T}} w_{\mathrm{T}} . H_{\mathrm{T}}$$

where H_T is the equivalent dose in the tissue or organ T and w_T is the weighting factor for tissue or organ T.

Table 3 The probability of radiation-induced stochastic effects, derived from the *1990 Recommendations of the ICRP* (ICRP 1991)

	Probability %/Sv			
Population	Fatal cancer	Non-fatal cancer	Severe hereditary effects	Total
Adult workers	4.0	0.8	0.8	5.6
Whole population	5.0	1.0	1.3	7.3

Table 4 Tissue weighting factors, w_T, as recommended in the *1990 Recommendations of the ICRP* (ICRP 1991)

Organ or tissue	w_T
Gonads	0.20
Bone marrow (red)	0.12
Colon	0.12
Lung	0.12
Stomach	0.12
Bladder	0.05
Breast	0.05
Liver	0.05
Oesophagus	0.05
Thyroid	0.05
Skin	0.01
Bone surface	0.01
Remainder	0.05

Based on the probabilities of stochastic effects and of deterministic effects, the ICRP in its 1990 Recommendations (ICRP 1991) recommended the new dose limits for workers and for the general public which are listed in Table 5. For workers the ICRP now recommends a limit of 20 mSv per year averaged over five years (100 mSv in 5 years), with the provision that the effective dose should not exceed 50 mSv in any single year (ICRP 1977). It will be seen from Table 5 that lower doses are recommended for members of the public. The limits for the occupational exposure of workers apply to men and women; however, should a woman become pregnant, it is necessary to provide protection for both the woman and her unborn child. Thus it is recommended that once pregnancy has been declared, the equivalent dose limit to the surface of the mother's abdomen should not exceed 2 mSv during the remainder of the pregnancy and that intakes of radionuclides should not exceed 1/20 of the Annual Limit on Intake (ALI).

Table 5 Dose limits recommended in the *1990 Recommendations of ICRP* (ICRP 1991)

	Dose limit	
Application	Worker	General public
Effective dose	20 mSv per year averaged over defined periods of five years	1 mSv in one year
Annual equivalent dose in:		
Lens of the eye	150 mSv	15 mSv
Skin	500 mSv	50 mSv
Hands and feet	500 mSv	–

For the radiation protection of persons working with radionuclides it is possible to calculate the maximum amount of a radionuclide compound which, if accumulated and retained in the body, will deliver the dose limit to the organs and tissues. This can be derived from the radiation characteristics of the nuclide and its biological behaviour. From these values the maximum amounts that could be taken into the human body in any year without exceeding the annual radiation dose limit may be calculated. These derived ALI vary widely from radionuclide to radionuclide. For example, for the low radiotoxicity nuclide ^{3}H, ingested as tritiated water, the ALI is 1.3 GBq; while for the highly radiotoxic, long-lived α-particle emitting nuclide ^{226}Ra, which deposits predominantly in the skeleton where it remains for many years, the ALI for ingestion is only 90 kBq.

Basic principles of radiation work

The basic objective of radiation protection is to minimize, or prevent, injury to individuals and to the general population. In order to achieve this objective it is necessary to ensure that all work involving radiation sources or radioactive materials is carried out carefully and with strict observance of a few basic principles. In particular, the activity of the source used should be the lowest that is necessary to yield valid and meaningful information.

In work with sources of penetrating radiation, that is radiation which may irradiate the body from outside, three cardinal points should be observed.

- The operations with the exposed source should be carried out in the minimum time consistent with the efficient completion of the task, since the shorter the time of exposure the lower is the dose received by the operator.
- The operator should work at the largest practical distance from the source, since, in general, radiation intensity decreases by the square of the distance from the source, i.e. the dose rate at two metres is only one quarter that at one metre from the source. (There may be exceptions to this general rule if the physical size of the source is large and/or if the construction of the unit results in significant scattering of the radiation.)
- The operation should always be carried out from behind a suitable screen or shield. Shielding is necessary in order to reduce the intensity of the radiation to an acceptable working level. The amount and nature of the shielding material required will vary

according to the type and intensity of the source and, preferably, the shielding should surround the whole source. For high-energy X- or γ-rays, several inches of lead may be needed, whereas high-energy β-particles may be almost completely stopped by 1 cm of 'perspex', polycarbonate or similar material; low-energy β-particles and α-particles have so little penetrating power that they are totally absorbed by the walls of their container.

These principles of *time*, *distance* and *shielding* apply to all work with radiation sources whether they be X-ray machines, or radionuclides sealed in leak-proof containers (so-called *sealed sources*) or solutions of, or some other preparations of, a radioactive material not permanently sealed into a container, which are called open, or *unsealed sources*.

Work with unsealed sources requires further precautions of which the most important is the principle of *containment*. This makes it necessary that equipment and working procedures are so designed that the escape of radioactive material from the immediate work area into the rest of the laboratory, or into the environment, is prevented or at least minimized, so that the risks of workers inhaling or ingesting radioactive materials, and of general contamination, are minimized. The type of containment and shielding required for such work will depend on the nature of the work, the total amount of radioactivity involved and the radiotoxicity of the radionuclides concerned.

Radiation protection in the animal house

The major procedures involving the use of sources of ionizing radiations in animal houses, or with which animal house personnel may be required to assist, are:

- The exposure of animals to X-rays for radiographic purposes, or to acute doses of X- or γ-rays or to neutrons for radiobiological studies.
- Exposure of animals to continuous (chronic) irradiation with γ-rays from a sealed radionuclide source for radiobiological studies.
- Administration of radioactive compounds to animals, by single or multiple injections, by single oral doses, by continuous feeding, or by inhalation, for the purpose of investigating metabolic pathways, the biokinetics of the radioactive substances in the body or for studies of radio- or chemical toxicity.

Work with X- and γ-rays

The exposure of animals to acute doses of radiation from X-ray machines or particle generators should pose little problem from the radiation protection point of view, provided that the equipment used is properly designed, fitted with adequate safety interlocks and that it is well-maintained. Most larger X-ray facilities will be fixed installations with the machines being located in separate rooms fitted with safety interlocks to prevent the machine being operated with the room door open. Ideally the facility will also be equipped with video or other devices to permit the operator, who will be outside the room, to see that no-one remains in the machine room. All persons using such radiation facilities must have the nature and operation of the equipment and its safety devices explained to them before they use the equipment. In no circumstances should the safety devices be disconnected or otherwise inactivated. If small portable diagnostic X-ray machines are used in an animal house, it will be necessary for the operator to make sure that they and anyone assisting is standing well clear of the area of the X-ray beam before making the exposure.

When animals are exposed to continuous irradiation from a γ-ray source it will be necessary for animal house staff and others to enter the irradiation facility regularly in order to feed and water the animals, to clean the cages and to carry out scientific observations. This situation also requires that the staff concerned should be properly trained in the safety procedures before they are allowed to work in the area. Gamma ray sources used for continuous irradiation usually consist of sealed capsules containing many GBq of ^{137}Cs or ^{60}Co; such a source must be housed in a well-designed irradiator unit which permits the source to be moved into a well-shielded, 'safe' position before the irradiation chamber can be opened. Safety interlocks are essential and other devices should include clearly visible, preferably illuminated indicators to show whether the source is in the 'safe' or the 'exposure' position. Before anyone starts working in a room housing a continuous irradiation facility, the source should be returned to its 'safe' position and maintained there until all persons have left the chamber. With large irradiation units, even when the source is in the 'safe' position, the radiation dose in the immediate vicinity of the unit, though small, will be higher than the normal radiation background; thus any operations within the vicinity of the unit should be completed as rapidly as possible. Time-consuming tasks such as cage cleaning and detailed examination of animals should be carried out in another room. All staff working with X- or γ-ray facilities should follow the advice given in the local rules for radiation work (see below) concerning the need to wear a film badge, or other type of personal monitor, while working in or near the irradiation facility.

Provided that the simple precautions outlined above are observed and the safety devices on the machines are maintained in good working order, the exposure of animals to X- or γ-rays should not result in any person receiving a significant dose of radiation.

Work with radioactive materials

The administration of radioactive materials to experimental animals poses far more problems from the point of view of radiation protection than work with X- or γ-rays from external sources. However, if personnel are properly trained and practise due care and responsibility in carrying out the various procedures, radiation exposures should be no greater than for staff working with X- or γ-ray sources.

The majority of experiments in which radioactive materials are administered to animals involve the use of relatively small quantities, < 200 kBq, of low or medium toxicity radionuclides, for example ^{3}H, ^{14}C or ^{59}Fe; thus even with γ-ray emitting nuclides there is little risk of significant external irradiation of animal house personnel by the treated animals. The major problems arise because there is a risk of internal contamination, which may result from inhalation or ingestion of radionuclides adsorbed onto dust particles arising from bedding materials, inhalation of gaseous radioactive compounds exhaled by the animals, contamination of the skin by excreta, or from bites and scratches. All these risks can be reduced to negligible proportions by the good design of facilities and operating procedures, and the maintenance at all times of high standards of personal and animal house hygiene.

For experiments involving the use of small quantities of radioactivity, for example < 50 MBq of low or medium toxicity radionuclides, or < 5 MBq of highly radiotoxic nuclides, elaborate facilities are not necessary. A separate room, or rooms, should be set aside for the housing of animals given radioactive materials, and should be well-ventilated and constructed so that the walls, floors and bench surfaces are non-absorbent and easily cleaned and, if necessary, decontaminated, with sealed joints between flooring materials, and between walls and ceiling. Floors should be cleaned by wet mopping rather than dry sweeping in order to reduce the risk of producing inhalable dust particles.

If it is necessary to work with larger amounts of radioactivity, or to administer radionuclides continuously in the diet, more elaborate facilities may be required in order to ensure containment of the radioactive materials within a well-defined area and the removal of any gaseous products through appropriate filters or other devices. Often it will be sufficient to keep the animals inside a fume hood, but if highly toxic nuclides are used it may be desirable to house them in glove boxes.

Animals should be housed in easily cleaned cages constructed of plastic, stainless steel or glass, which are kept separate from the cages used for non-radioactive animals. Bedding should be non-dusty, such as cellulose wadding, absorbent paper or coarse grain vermiculite. When only small amounts of radioactivity are involved and excretion is expected to be low, a coarse sawdust may be used. Contaminated bedding, excreta, etc., will have to be disposed of as radioactive waste, as discussed later. If the work involves the use of larger animals such as sheep, dogs, or primates, the pens or cages in which the animals are housed should be so designed that they are easily cleaned and decontaminated by hosing down or mopping out. Depending on the amount of radioactivity involved, it may be necessary for the drains from these pens to lead into delay tanks so that the effluent may be treated or stored before final disposal.

In addition to the rooms set aside for housing radioactive animals, it is desirable to have one or more separate rooms for the preparation and administration of the radioactive materials, and for the cleaning of radioactive cages and equipment. Such rooms should be designed to allow easy cleaning, and if necessary decontaminating, and be equipped with dedicated equipment for, for example, preparing radioactively labelled diets, injection solutions or for weighing radioactive animals.

In the design of modern high-grade animal houses with facilities for work with radioactive materials, conflicts may arise between the requirements for good animal husbandry and infection control, and those for safe working with radionuclides. A common example is found in the requirements for ventilation systems: minimization of the risk of entry of microorganisms usually calls for the maintenance of a slightly higher pressure inside the animal rooms compared to surrounding areas; in contrast, in radioactive laboratories a slightly negative pressure is recommended in order to minimize the spread of airborne radioactivity. Special problems may also arise if it is necessary to install fume hoods or laminar air flow workstations inside animal rooms. In such situations some compromises have to be made and it is essential that the design of the facility is undertaken in close collaboration with experienced animal technicians, laboratory animal scientists, the radiation protection adviser, architects and heating and ventilation engineers.

As with infection control, the maintenance of a high standard of hygiene in the animal house is an essential part of good radiation protection practice. Cages and rooms should be cleaned regularly and thoroughly; only essential equipment should be stored in the rooms reserved for radioactive work and a high standard of tidiness should be maintained at all times.

Detailed protocols for all work with radioactive materials should be included in the local rules, which should also contain basic rules for laboratory and personal hygiene. The following are considered to be essential:

1) Rubber or plastic gloves should be worn at all times when working with radioactive materials and in areas where radioactive work is carried out. It should always be remembered that anything touched with contaminated gloves will usually become contaminated; consequently it is necessary to avoid touching switches, taps or apparatus outside the radiation area with gloved hands. Gloves should be cleaned while still on the hands, and the hands washed thoroughly and monitored for radioactivity immediately after leaving the area where radioactive work was carried out. Whenever possible disposable gloves should be used.
2) Separate overalls, or other protective clothing, should be kept for work in the areas where radioactive work is carried out and they should not be worn outside that area.
3) For any work involving the risk of inhalation of dust, for example from cage cleaning, an appropriate, i.e. specifically approved for the purpose, well-fitting dust mask or respirator which completely covers the nose and mouth should be worn.
4) When working with radioactive materials, bench tops should be covered with absorbent paper or other disposable material. Fluids for injection should be placed in trays lined with absorbent paper, so that any spillage is contained within the tray. Bottles or flasks containing the radioactive solution should also be placed in holders that provide any necessary shielding and also minimize the risk of them being knocked over by accident.
5) No radioactive solution, or any other solution used in the area in which radioactive work is carried out should be drawn into a pipette by mouth; suitable pipetting devices should always be used.
6) Eating, drinking, smoking, the taking of medication or any other form of drug, and the application of cosmetics are absolutely forbidden in areas where radioactive work is carried out. Obviously, it is also forbidden to store foodstuffs, personal drinking vessels or smoking materials in such areas.

Inside areas where radioactive work is carried out, and in their immediate vicinity, a regular programme of monitoring for radioactive contamination should be carried out, and suitable monitoring equipment should be an essential part of the equipment for such areas. Monitoring should include floors, walls, benches, ceilings, sinks, waste traps, as well as cages and other equipment. Staff should be taught as a routine to monitor their hands, feet and clothing at the end of each session of work in the radioactive area. Regulation 24 of the Ionising Radiations Regulations 1985 (Statutory Instruments 1985) makes monitoring and the provision of appropriate equipment a legal requirement; the UK Approved Code of Practice gives further advice on the implementation of this regulation. Monitoring of hands and clothing for low-energy β-particle emitting radionuclides such as ^{3}H and ^{14}C is impracticable, and consequently the very high standards of personal hygiene that are necessary for all work with radioactive materials become of even more critical importance when working with these nuclides. 'Biological' monitoring of personnel for internal contamination with ^{3}H and ^{14}C by measurement of the nuclides in samples of urine is possible, and in some circumstances it may be considered desirable to conduct such monitoring on a routine basis.

This section has been concerned only with discussing the basic principles that must be followed for work with radiation and radioactive materials in animal houses. The detailed protocols for such work and the specific local requirements should be incorporated into the local rules with which every person working with radiation sources of any kind should be fully conversant. Detailed advice on local matters can be obtained from the local Radiation Protection Supervisor, who is the person appointed by the employer to ensure that radiation work is carried out in compliance with the Ionising Radiations Regulations 1985 (Statutory Instruments 1985). More detailed general information can be obtained by reference to some of the works listed in the references. In the UK, advice on most matters relating to radiation protection can be obtained from the National Radiological Protection Board (NRPB), Chilton, Didcot, Oxfordshire OX11 0RQ.

Protection against ultraviolet and microwave radiation

In contrast to the comprehensive international recommendations, and national or international regulations, codes of practice and guidance for the conduct of work with ionizing radiations, there is at present little authoritative guidance on work with non-ionizing radiation such as UV and microwave radiation. An International Commission on Non-Ionising Radiation Protection (ICNIRP) (Bernhardt & Matthes 1997) was formed in 1992 and guidelines for the protection of workers against non-ionizing radiations are

Table 6 Basic restrictions on human exposure to electromagnetic radiation (frequency range 100 kHz to 300 GHz) (NRPB 1993)

Frequency	Basic restriction*	Comments
100 kHz–10 MHz	$0.4\,W\,kg^{-1}$**	SAR‡ averaged over whole body
	$10\,W\,kg^{-1}$[10 g]***	SAR in head and fetus
	$10\,W\,kg^{-1}$ [100 g]***	SAR in neck and trunk
	$20\,W\,kg^{-1}$ [100 g]***	SAR in limbs
	$f(Hz)/100mA\ m^{-2}$	Current density in head, neck and trunk
10 Mhz–10 GHz	$0.410\,W\,kg^{-1}$ **	SAR averaged over whole body
	$10\,W\,kg^{-1}$ [10 g]***	SAR in head and fetus
	$10\,W\,kg^{-1}$ [100 g]***	SAR in neck and trunk
	$20\,W\,kg^{-1}$ [100 g]***	SAR in limbs
10 GHz–300 GHz	$100\,W\,m^{-2}$†	Power density on any part of body

* Averaged over the masses shown in parentheses
** Averaged over any period of 15 minutes
*** Averaged over any period of 6 minutes
† Averaged over $68/f^{1.05}$ minutes, where f is the frequency in GHz
‡ SAR = Specific Absorption Rate

now published (ICNIRP 1996). In the UK the NRPB provides guidance and advice on the hazards of UV, microwave and other types of non-ionizing radiation, and recommends what are termed 'basic restrictions' on exposure to electromagnetic radiations. These 'basic restrictions', which were revised in 1993 (NRPB 1993), are based on biological considerations and are summarized in Table 6. NRPB also offers short training courses on protection against non-ionizing radiations.

The principal source of UV radiation in animal houses is probably in the use of so-called germicidal lamps for increasing protection against infection in some parts of the facility, although some experimental studies of the effects of UV, or laser radiation on living animals may be carried out in some facilities. Germicidal lamps emit almost monochromatic radiation at 254 nm, while lasers and other sources may be tuned to emit radiation at selected wavelengths within the UV spectrum (100–400 nm).

The eye is the critical organ for UV radiation and must be protected by suitable UV-absorbing goggles (Elder 1973). When the eyes are protected, the exposed skin may become the critical organ. Skin exposures to 254 nm UV radiation at up to 0.5 $\mu W\,cm^{-2}$ throughout an eight-hour working day appear to cause no problems. Somewhat higher exposures at higher or lower wavelengths may be permissible since UV radiation at 254 nm appears to be the most effective for producing skin damage.

In all work with UV radiation, care should be taken to avoid direct irradiation of the eyes or any other part of the body from the source. Therefore, if it is necessary to work for more that a few minutes in an area protected by a germicidal lamp, the lamp should be switched off for that period. International guidelines on UV radiation exposure limits have now been published (ICNIRP 1996).

The eyes and skin are also the critical organs for microwave radiation and every care should be taken to avoid direct exposures to microwave sources. International guidelines on limits for exposure to microwaves have been published (International Non-Ionizing Radiation Committee 1988). In the UK the NRPB has recommended a 'basic restriction' for exposure to $0.4\,W\,kg^{-1}$ to prevent thermally based effects in people (Table 6); while in the United States of America a limit of $10\,mW\,cm^{-2}$ has been proposed.

Until such time as clearer guidance is available, it is advisable in all work with microwave sources to design the facilities and procedures in such a way that the operator and others in the area are protected from both direct and indirect radiation. In designing working protocols, the advice of a safety officer with experience of such sources should be sought.

Regulatory aspects of radiation work

In most of the developed countries of the world all use of X-ray machines, radioactive materials and other radiation sources is controlled by some form of legislation. In most countries this legislation, and the associated regulations, follow closely the philosophy and dose limitations recommended by the ICRP. The latest, 1990 Recommendations of the ICRP, which involve some changes in the basic philosophy and major changes in the numerical values of the dose limits, have been discussed in a previous section; these changes will, eventually, lead to alterations of the legislation and its associated regulations.

All Member States of the European Community, including the United Kingdom, are subject to the provisions of the Treaty which created the European Atomic Energy Community (Euratom). Article 30 of this Treaty requires that basic standards be formulated for the protection of the general public and of workers against the dangers of ionizing radiations. Under Article 33 of the Treaty, Member States are required to enact the appropriate legislation and regulations to enable them to comply with these standards which have been laid down in the European Council Directives. The latest Directive, 96/29-Euratom, which takes account of the new 1990 Recommendations of the ICRP, came into force in May 1996. Until such time as national laws have been modified, the dose limits specified in the relevant national Regulations still have the force of law. However, for practical radiation protection purposes it would be wise to keep exposures below the limits recommended in the 1990 ICRP Recommendations; in some countries official guidance has already been issued on this matter. For example, in the UK the NRPB (Clarke 1993) has advised an occupational dose limit of

20 mSv in any single year, and that for pregnant women exposure should be as low as reasonably achievable such as to make it unlikely that the equivalent dose to the fetus will exceed 1 mSv during the remainder of the pregnancy (NRPB 1993). New UK Regulations are expected to be published in 1999–2000 (see below).

As a result of the Euratom Treaty and the resulting Directives as well as national legislation, radiation protection standards and regulations are broadly similar throughout most countries of Western Europe. At the international level, the International Atomic Energy Agency (IAEA) publishes *Basic Safety Standards for Radiation Protection*, the most recent edition of which was published in 1996 (IAEA 1996). These basic safety standards provide internationally harmonized guidance on radiation protection standards.

In the United Kingdom the Health and Safety at Work etc. Act 1974 governs all aspects of health and safety in the workplace, including work with radiation and radioactive materials. Radiation work is covered specifically by *The Ionising Radiations Regulations 1985* (Statutory Instruments 1985) and its associated Approved Code of Practice, *The protection of persons against ionising radiation arising from any work activity* (Health and Safety Commission 1985). (New Regulations are expected to be published in December 1999 and to come into force on 1 January 2000.) The Regulations have the force of law and cover all aspects of radiation work including health surveillance, dose limits, designation of controlled and supervised work areas, training of personnel, assessment of hazards, emergency planning etc.; the Approved Code of Practice provides guidance on the implementation of the Regulations. Section 2 of Part 2 of the Code of Practice which deals with non-medical radiography and irradiation, including research, is especially relevant to radiation work in laboratories and animal houses.

The basic responsibility for radiation protection lies with the employing organization, and Regulation 10 of the Code of Practice requires that employers appoint a suitably qualified person as Radiation Protection Adviser (RPA) to advise on all aspects of radiation protection. Regulation 11 also calls for the appointment of Radiation Protection Supervisors (RPS) who should be trained and competent to assist the employer in ensuring that radiation work complies with the requirements of the Regulations. Regulation 11 also provides for the preparation of written local rules for the conduct of radiation work. These local rules must contain descriptions of the controlled and supervised areas; written systems of working procedures, especially for non-classified persons who enter controlled areas; procedures for restricting access to controlled areas; and contingency plans for dealing with accidents or emergencies. It is important that all persons working with radiation or radioactive materials should know and understand the local rules, and should know how to contact their RPS should they need advice or should an emergency occur.

Work with radioactive materials is also controlled by the Radioactive Substances Act 1993, which replaced the older Acts of 1948 and 1960. This act covers the registration and authorization of the use of radioactive materials, their holding and storage and the licensing and disposal of radioactive waste.

National legislation controlling radiation in other countries is broadly similar to that in the UK, although the laws may differ in points of detail and administration. For example, in Germany work is controlled under the Atomgesetze and the Strahlenschutzverordnungen (Federal Republic of Germany 1989) and the Röntgenverordnungen (Federal Republic of Germany 1986). In the United States of America the control of work with reactor-produced radionuclides is governed by the United States of America Atomic Energy Act of 1954, while work with other radiation sources, including X-rays, microwaves and other radiation from electronic devices is controlled under the United States of America Radiation Control for Health and Safety Act of 1968. The detailed control of radioactive materials and radiation sources is laid down in the United States of America Code of Federal Regulations, Title 10 (10CFR), 1992, especially Parts 20, 30, 31, and 35, while transport is covered by 10CFR49. As in Europe, registration of radiation work, the appointment of RPAs or RPSs and the licensing of radioactive waste disposal are required.

In all legitimate work with radiation sources or radioactive materials in any country, it should not be forgotten that while the ultimate legal responsibility for radiation protection always lies with the employing organization, in practice the *primary* responsibility for protecting him, or herself, and other people from radiation hazards resulting from their own work lies with the individual worker. Consequently no one should be permitted to begin to use radiation sources in any form until they have become fully acquainted with the nature of the potential hazards and the general and local rules and procedures for such work. In the UK formal records of training and competence must be kept (Health and Safety Commission 1992) and broadly similar certification of training and competence are required in many other countries.

Disposal of radioactive waste

The disposal of radioactive waste is now strictly controlled in an increasing number of countries and, because of public concern about perceived or real hazards, its 'safe' disposal is becoming a difficult and expensive problem. In the UK, under the Radioactive Substances Act 1993, the control of radioactive waste disposal is vested in Her Majesty's Inspectorate of Pollution, which is responsible for registering all producers of radioactive waste and licensing disposal routes.

The principal type of radioactive waste produced in animal houses is likely to be the so-called 'low-level waste', which is defined as waste containing radioactive materials other than those acceptable for dustbin disposal, but not exceeding 4 GBq/tonne of α-activity or 12 GBq/tonne of β- or γ-activity. This low-level waste from an animal facility will consist of items such as plastic gloves, disposable syringes, disposable or contaminated clothing, animal bedding and animal carcasses.

The exact methods by which radioactive waste may be legitimately disposed of will depend on the terms and conditions of the authorization granted by the relevant authority, e.g. HM Inspectorate of Pollution in the UK. These conditions will vary according to the particular circumstances, and no general advice will be given here. In some situations if the activity to be disposed of is very low, less than a few kBq m^{-3}, disposal via the normal domestic waste disposal system, or local incineration, may be permitted. Similarly, disposal of very low-level liquid waste via the normal sewage system may be permitted. In all other circumstances disposal of the radioactive waste will generally take the form of delivery to an authorized disposal or burial site, by an organization licensed to transport such materials.

Advice on the permitted method of disposal of radioactive waste should always be sought from the local RPS before attempting to dispose of any contaminated, or potentially contaminated item.

Emergency procedures

The UK Approved Code of Practice requires, as do similar regulations in other countries and the IAEA Basic Safety Standards (1996), that organizations prepare contingency plans to deal with accidents or unplanned occurrences during work with radiation sources or the transport of radioactive materials; such plans must be written down in the local rules.

In all situations where radioactive solutions or other open sources of radioactivity are used, it is essential that clear guidance be given in the local rules regarding the action to be taken in the event of a significant spillage of radioactivity. It is also desirable that all the equipment likely to be needed for dealing with a spillage of radioactivity, including the appropriate monitoring devices, should be kept available at all times and that all staff should know how to use the emergency equipment and where it is to be found.

While in most animal facilities the likelihood of a serious radiation accident or a large spillage of radioactivity is remote, it is essential that all staff are instructed regularly in the procedures to be adopted should an emergency occur.

Conclusions

Although the use, and especially the misuse, of all sources of ionizing and non-ionizing radiations is potentially capable of causing damage – sometimes very serious damage – to human health, the risks are small if both the employing organization and the individual worker adopt a thoroughly responsible attitude to radiation work. The ultimate legal responsibility for the safe conduct of radiation work lies with the employer; nevertheless, individual workers do have a primary responsibility to protect themselves and their colleagues from any potentially harmful effects of radiation that may arise from their personal activities in the workplace. Thus it is essential that before starting work with radiation sources or with radioactive materials all workers are fully instructed in, and understand, the nature of the risks and the reasons underlying the safety procedures and the working methods they are required to practise.

Safe and trouble-free work with any source of ionizing, or non-ionizing radiation, or indeed with any other hazardous agent, requires the individual to exercise care, self-discipline and above all common-sense at all times. Familiarity with radioactive materials or other radiation sources should never be allowed to develop into neglect of, or contempt for, their potential dangers and, like electricity, fire and all other types of hazardous agent, radiation should be treated with respect but without fear.

References

Arena V (1971) *Ionising radiation and life*. London: Kimpton

Bernhardt JH, Matthes R (1997) Recent and future activities of ICNIRP. *Radiation Protection Dosimetry* **72**, 167–76

Clarke RH (1993) Response by NRPB to the 1990 Recommendations of ICRP. *NRPB Radiological Protection Bulletin* No. 141, 4–8

Elder RL (1973) Lasers and eye protection. *Science* **182**, 1080

European Council Directive 96/29-Euratom (1996) Laying down basic safety standards for the protection of the health of workers and the general public against the dangers arising from ionising radiation 13 May 1996. *Official Journal of the European Communities* L159, 29 June 1996

Federal Republic of Germany, Röntgenverordnungen (1986) *Bundesgesetzblatt* 923, 1989

Federal Republic of Germany (1989) Neufassung der Strahlenschutzverordnung, *Bundesgesetzblatt* 1321, 1986

Health and Safety at Work etc. Act(1974). London: HMSO

Health and Safety Commission (1985) *Approved Code of Practice – The protection of persons against ionising radiation arising from any work activity.* London: HMSO

Health and Safety Commission (1992) *Management of health and safety at work regulations: Approved code of practice.* London: HMSO

Hughes S (1993) Radiation exposure in the UK. *NRPB Radiological Protection Bulletin* No 145, 10–12

International Atomic Energy Agency (1996) *Basic Safety Standards for Radiation Protection (1996 Edition), Safety Series IAEA.* Vienna: IAEA

International Commission on Non-Ionizing Radiation Protection (1996) Guidelines on UV Exposure Limits. *Health Physics* **71**, 978

International Commission on Radiological Protection, Publication 26 (1977) Recommendations of the International Commission on Radiological Protection. *Annals of the ICRP* **1** (3)

International Commission on Radiological Protection, Publication 60 (1991) 1990 Recommendations of the International Commission on Radiological Protection. *Annals of the ICRP* **21**(1–3)

International Non-ionizing Radiation Committee/International Radiation Protection Association (1988) Guidelines on limits of exposure to radiofrequency electromagnetic fields in the frequency range from 100 KHz to 300 GHz. *Health Physics* **54**, 115–23

NRPB (1993) Board Statement on restrictions on human exposure to static and time varying electromagnetic fields and radiation. *Doc. NRPB* **4**, No 5, 1

Radioactive Substances Act (1948) London: HMSO

Radioactive Substances Act (1960) London: HMSO

Radioactive Substances Act (1993) London: HMSO

Statutory Instruments (1985) No. 1333. *The Ionising Radiations Regulations 1985.* London: HMSO

United States of America Atomic Energy Act (1954) Washington DC: US Government Printing Office

United States of America Radiation Control for Health and Safety Act (1968) Washington DC: US Government Printing Office

United States of America Code of Federal Regulations Title 10 (1992) Washington DC: US Government Printing Office

Further reading

Baranski S, Czerski P (1976) *Biological Effects of Microwaves.* Stroudbirg, PA: Dowden

Dennis JA, Stather J (1997) Non-ionising radiation. Special Issue of *Radiation Protection Dosimetry* **72** (3/4)

National Radiological Protection Board (1989) *Living with Radiation.* London: HMSO

Noz ME, Maguire GQ (1995) *Radiation Protection in the Health Sciences.* Singapore: World Scientific

Shapiro J (1972) *Radiation Protection: a Guide for Scientists.* Cambridge, Mass: Harvard University Press

Sumner D, Wheldon T, Watson W (1991) *Radiation Risks.* 3rd edn. Newton Stewart: Tarragon Press

Sowby FD (1965) Radiation and other risks. *Health Physics* **11**, 879

7

Safety management

John Ryder

Contents

Introduction

Safety is important. It is important enough to be part of the day-to-day responsibilities of all managers and staff. Although most of this chapter is devoted to the actions of individuals with specific safety responsibilities, we must not lose sight of the following principles:

- all managers are responsible for ensuring that the activities they manage are carried out safely, as well as efficiently;
- all staff are responsible for working in a safe manner, and informing their managers of any safety problems that arise.

So if safety is everyone's responsibility, why do we need safety management? There are two main reasons:

- safety is a complex subject, given the plethora of hazards and legal requirements, hence managers and staff need to have someone that they can turn to for comprehensive guidance on current thinking and requirements;
- we live in an increasingly competitive world, in which we are all under pressure to achieve more with finite resources, hence there is a danger that safety will be marginalized if it does not have a champion to keep its profile high.

Safe working requires good management by people who care about safety and are empowered to do the things needed to achieve it. Buildings and equipment of a suitable standard and a well-informed and well-trained workforce are essential.

HSE guidance on health and safety management (Health and Safety Executive 1997, 1998) describes five steps to reducing accidents and work-related illness:

- draw up an effective health and safety policy;
- organize staff to ensure competence, control, cooperation and communication;
- identify objectives and set performance standards;
- measure performance against these standards;
- audit systems, review performance and learn from experience.

Setting up a safety management structure

The Management of Health and Safety at Work Regulations 1992 (Statutory Instruments 1992a) require every employer to 'appoint one or more competent persons to assist him in undertaking the measures he needs to take to comply with the requirements and prohibitions imposed upon him by or under the relevant statutory provisions'. The Approved Code of Practice (Health and Safety Commission 1992) which accompanies these regulations makes it clear that these competent persons may be employees or outside consultants or a mixture of both. There is no requirement for any appointment to be full-time. Competence does not require the possession of a specific qualification, but must encompass an understanding of relevant current best practice, an awareness of the limitations of their own experience and knowledge, and the willingness and ability to supplement their existing experience and knowledge as the need arises.

Given the speed with which safety issues can arise, and the urgency with which they may need to be addressed, most employers will wish to safeguard their positions by appointing at least one on-site employee as a competent person in safety matters. For the remainder of this chapter this employee will be referred to as the Safety Officer. In some organizations the term Safety Adviser is used to emphasize the role of the person in advising and supporting line management, which retains the primary responsibility for safety. In a very small organization, the role of Safety Officer may represent as little as 10% of one person's time. In this case, one would look for a person with several years' experience of the job in hand, such as a senior animal technician, and an interest in safety. These basic requisites would be supplemented by attendance at relevant meetings and short courses, and by access to outside help when needed. At the other extreme in a large organization, the Safety Officer will be a full-time appointment supported by a staff of specialists in various aspects of occupational health and occupational hygiene. Such a Safety Officer and specialists would require qualifications and experience sufficient to gain membership of an appropriate professional body such as the Institute of Occupational Safety and Health (IOSH).

The role of the Safety Officer is generally fourfold:

- to inform the employer of legal requirements and best practice in health and safety;
- to ensure compliance with legal requirements and best practice. The Safety Officer may delegate actions, but must always retain an up-to-date overview;

- to investigate emerging health and safety problems, and take or recommend appropriate action;
- to act as a repository of knowledge and wisdom, accessible to all managers and other staff.

The employer should involve the Safety Officer and safety specialists to liaise with appropriate managers in the planning and design of new or refurbished facilities from the initial concept. This will help to ensure that safety requirements are built in, avoiding the need for expensive additions at a later date.

It is clear that the Safety Officer must possess maturity, good judgement and good interpersonal skills, especially persuasiveness. From time to time he or she will have to give advice that the recipient does not want to hear!

It is essential that safety has a voice at the highest level in an organization, in order that safety requirements are not overlooked when resources are allocated and policies formulated. The Safety Officer may be able to fulfil this function, but in many organizations will not be senior enough in the hierarchy to wield the required influence. In this case, a senior manager in the organization with ready access to the Managing Director, Chief Executive Officer or equivalent, should take on the role of safety champion. The safety champion will need regular contact with the Safety Officer to maintain an awareness of important issues, especially those that have resource implications or require policy decisions.

Any organization, unless very small, should have a Safety Committee, whether or not it is a legal requirement (Statutory Instruments 1977), as it serves a useful purpose. The Safety Committee acts as a forum for discussion of safety issues and an alternative route, other than line management, by which staff can bring problems to the attention of the employer. The Safety Officer will be a member of the Committee, and in many cases will chair it. Depending on the constitution of the Committee, the members may be elected or appointed. In either case, they should have a reasonably detailed knowledge of the areas they represent and a genuine interest in safety. They may be given specific responsibilities for safety actions in their own departments, e.g. for induction training of new staff. In practice, many of the items actually discussed at Safety Committee meetings are concerned with equipment maintenance, selection of protective clothing and safety-related policies administered by the personnel department. It is helpful to have a representative from maintenance, purchasing or personnel present when such items are discussed.

Safety, GLP and quality assurance

Safety overlaps with Good Laboratory Practice (GLP) and Quality Assurance (QA). GLP and associated QA requirements apply formally to laboratory animal facilities that generate safety data for purposes of product registration (Statutory Instruments 1997). GLP imposes strict requirements for documentation, authorization of work, and adherence to written procedures. QA provides a mechanism for checking compliance with these requirements. Thus, in a GLP environment, there will be a mechanism for authorizing, issuing and updating Standard Operating Procedures (SOPs), and there will be a QA Unit. In such an environment, it is convenient and efficient to document specific safety requirements through SOPs, rather than to set up a separate system of Codes of Practice. Similarly, it is convenient to have the QA unit check and report on all aspects of compliance with SOPs, including compliance with safety requirements. This also helps to reinforce the point that safety is an integral part of the correct performance of any task, not a bolt-on extra. Such principles of controlled experimentation encompassed by GLP are applicable to any biomedical research laboratory.

Basic documentation

Policy

The Health and Safety at Work etc. Act 1974 requires 'every employer to prepare and as often as may be appropriate revise a written statement of his general policy with respect to the health and safety at work of his employees and the organisation and arrangements for the time being in force for carrying out that policy, and to bring the statement and any revision of it to the notice of all of his employees'. For very small organizations with few hazards, it may be possible to put all the written safety instructions to staff in this one document. For larger organizations, with a variety of hazards, this approach would soon become unwieldy. A tiered approach, which presents a general statement to all staff and specific detailed instructions to staff working in individual areas, is more practicable. The general statement might be written by the Safety Officer; the specific detailed instructions might be drafted by someone with a detailed working knowledge of the area concerned, and approved by the Safety Officer or by a senior manager in the area.

The author has assisted one company to adopt a two-tier approach:

- a Safety Manual contains a statement of company safety policy, including a general description of the principal hazards present and safety precautions required, a description of the company's health and safety organization, with the names and internal telephone numbers of all relevant individuals, and emergency procedures. Copies of the Safety Manual are held by all managers and members of the Safety Committee, and are available to all staff for reference. An extract from the Safety Manual appears in the Staff Handbook which is issued to all employees.
- Safety SOPs describe in detail the precautions to be taken in individual areas and identify, by job title, the staff who are responsible for implementing them. The Safety SOPs for the animal facilities also cover entry restrictions, hygiene requirements, arrangements for secure storage of controlled drugs, and vaccination of staff. Each Safety SOP is prepared by a senior manager familiar with the area concerned.

Regardless of the form of documentation used, it is essential to have a mechanism for updating it when necessary.

ESAC has published guidance on safety policies in the education sector (HSC Education Services Advisory Committee 1985).

Information and training

The word 'information' occurs frequently in regulations issued in the last few years, e.g. in the Control of Substances Hazardous to Health Regulations (COSHH) (Statutory Instruments 1999). It is no longer enough to provide staff with the necessary equipment, protective clothing and training in how to do the job, although all these things continue to be necessary. We must also inform them of the hazards and risks associated with their work, so that they understand why they must take the specified precautions, and not just follow them by rote (or when their supervisor is watching).

Sooner or later, the need to provide information comes into conflict with the need to maintain confidentiality. Some things may need to be kept confidential, e.g. the intended use of a novel material that is being tested, or clinical information about an individual, but beware the temptation to use 'confidentiality' as an excuse to restrict information that is merely inconvenient or embarrassing. Treat information as confidential only when necessary and for clearly understood reasons. Thus you should always be prepared to discuss with your Safety Committee accidents that have occurred, or symptoms that have developed, even though you may need to withhold the identity of the person concerned.

Safety training is a requirement for all staff who work in areas where hazards are present. In effect this means everyone, since no areas of a workplace are likely to be completely free from hazards. The amount of training required will depend on the number of hazards and the magnitude of the risks. By general workplace standards, animal houses are usually medium risk, i.e. the risks are greater than in an office, but less than on a building site.

The first, and possibly most important, component of safety training is induction training, which should be given to each newly recruited person, including temporary staff and those on training schemes and work placements, on their first morning at work. At its most basic, the purpose of this training is to protect new employees from injury or illness during their first few weeks of employment, when they will not be fully familiar with their new working environment, and may be at their most vulnerable. It should emphasize basic items such as what to do if the fire alarm rings, areas which must not be entered or animals or chemicals which must not be approached until training has been given, what to do if you find electrical equipment with a damaged cable and how to get first aid in the event of an accident.

More comprehensive training must be offered later. In general, safety should be taught not as a separate subject but as an integral part of how to do a job. This applies whether training is on-the-job, by attendance at in-house courses, or at college. The exception is for those who are to have specific safety responsibilities for which qualifications, such as the National Certificate or National Diploma in Occupational Safety and Health, awarded by the National Examination Board in Occupational Safety and Health (NEBOSH), or attendance at some of the Royal Society for the Prevention of Accidents (RoSPA) safety courses, are likely to be appropriate.

A variety of training videos is available covering specific hazards found at work. Often, a video is accompanied by a trainer's manual which provides information in greater depth, and gives the answers to questions that may be asked in a training session. Other training aids include tape/slide presentations, and low-cost, cartoon-style booklets which can be given out to all relevant staff.

Training records

It is recommended that all organizations should have training records for each employee showing the training and instruction received and the techniques which the person is competent to perform. In a GLP environment, such training records are compulsory. The training records should include details of safety training, including the induction training described above.

Accident reports

Every organization should have a system for recording and investigating accidents and other dangerous occurrences. Accident reports are of two types:

- Reports internal to the organization. A report should be made for every accident that causes an injury, however trivial it appears to be, and every dangerous occurrence. The report should be made in the Accident Book (Form B1 510), as required by the Social Security Act 1975. The Accident Book should be freely available to all staff. The report should include at least the full name and occupation of the person involved, details of any injury sustained, the date and time, location and description of the accident or occurrence, including any animals, chemicals or machinery involved, any first aid given and any follow-up actions taken. The Safety Officer should receive a copy of each report, and should present a summary of the reports received to the Safety Committee for discussion. In larger organizations, it may be convenient to enter each report into a computerized database, so that summary information and trends can be extracted readily. An HSE booklet (Health and Safety Executive 1991) gives advice on how to extract useful information from accident reports.
- Reports that the organization is required by the RIDDOR Regulations (Statutory Instruments 1995) to make to the Health and Safety Executive or local authority. These are required only for more serious accidents, such as those resulting in death, major injury, or more than three days absence from work, for specified dangerous occurrences and for reportable diseases. An HSE booklet (Health and Safety Executive 1996) gives guidance. To ensure a uniformity of approach, it may be that reporting to the HSE or the local authority should normally be done by the Safety Officer. It should be noted, however, that deaths, major injuries and dangerous occurrences must be notified to the HSE 'by the quickest practicable means', i.e. by telephone. Suitable arrangements must be in place to do this if the Safety Officer is absent.

Various other records are described in later sections of this chapter.

Responding to problems

It is not enough for an organization to have systems for recording and summarizing health and safety problems. Effective action must be taken to investigate and remedy them. In practice, problems requiring action are likely to be of the following types:

Accidents, reported as above. The most frequent types of accident in an animal facility are likely to be animal bites, minor cuts and bruises and chemical splashes. The occasional needlestick injury is also to be expected.

Illnesses. Information on these tends to surface by a variety of routes, e.g. accident report forms, self-certification of sickness absence, medical certificates, direct reports to supervisors or to the personnel department, or from health surveillance. Look out for reports of nausea (which may be a consequence of exposure to chemicals), skin rashes, watery or itchy eyes, tightness of the chest, wheezing or asthma (which may be due to laboratory animal allergy or exposure to chemicals), and back problems (which may be due to poor techniques of handling animals or cages).

Complaints and criticisms from staff. These may relate to illnesses, as above, or to conditions which staff perceive to be unsafe, e.g. chemical smells, damaged electrical cables, inadequate arrangements for disposal of syringes and needles.

Complaints and criticisms from neighbours. These may relate to pollution of the environment, e.g. by smoke emission from incinerator chimneys, or the appearance of hazardous waste such as chemicals or used syringes in places accessible to the public.

Criticisms from HSE inspectors. HSE inspectors may visit at any time, with minimal or no notice. Such a visit may be triggered by the reporting of an accident or dangerous occurrence, or may be routine. Serious deviations from acceptable standards may result in an enforcement notice, a prohibition notice, or prosecution. In reasonably well run establishments, a more likely outcome to the discovery of faults is that the visit will be followed by a letter recommending improvements.

Minor accidents and complaints from staff may be investigated by local management and the appropriate member of the Safety

Committee, but for more serious problems the Safety Officer must be involved and a panel of staff may be appointed to investigate. Cases of work-related illness may be investigated by the Safety Officer, but if the organization has an occupational health unit or a medical adviser, they should be brought in. The maintenance department may need to test equipment. A complaint from a member of the public will often require the involvement of the waste-disposal section. Letters from the HSE must never be ignored: recommendations should be acted on, or a reasoned response should be made if an alternative course of action is thought to be better.

Most health and safety problems arise from one or more of the following factors:

- faulty technique
- inadequate facilities
- faulty or inadequate equipment (including personal protective equipment)
- failure to follow correct procedures
- inadequate information on hazards.

Establishing the cause of a problem often suggests a solution. Faulty technique may be correctable by further training. Failure to follow correct procedures may be corrected by further training and instruction, accompanied in extreme cases by disciplinary action. Inadequate facilities and equipment require replacement or improvement. Faulty equipment requires repair or replacement. Unfortunately, these actions may compete for scarce financial resources, which is where the safety champion must be brought in. If the required improvements cannot be made, then the only alternative will be to restrict the range of activities to those that can be performed safely.

In some cases, the best way to proceed is to repeat the risk assessment (see next section) since the occurrence of an unexpected problem may indicate that the original assessment is no longer valid. If available information is inadequate, the remedy may be to conduct a literature search, or to chase suppliers for data sheets.

When responding to problems, it should always be remembered that the objectives are twofold:

- to remove the hazard or reduce the real risk;
- to restore confidence in the adequacy of the safety precautions.

The first objective will test the knowledge and ingenuity of staff and their managers in the area, supported by the Safety Officer; the second will test interpersonal and communication skills. It is not enough to be right, you also have to be believed!

Organizing risk assessment

The Management of Health and Safety at Work Regulations 1992 (Statutory Instruments 1992a), which implement the European Framework Directive (Council Directives 1989), require every employer to make a suitable and sufficient assessment of all risks to health and safety, not just those covered by more specific legislation. These regulations provide a sensible starting point for the Safety Officer who is trying to comply with all applicable safety legislation. They require an overview of all hazards present in the workplace, the risk (probability and severity) of harm which may result from each hazard, and identification of statutory requirements which must be satisfied. The hazards likely to be found in the animal house, and the relevant legislation, are the subjects of other chapters in this book: Table 1 provides a brief summary and identifies the European Directives that have been implemented in the UK by Regulations.

Who should perform the initial overview required by the Management of Health and Safety at Work Regulations? Because it is an overview, it is probably best done by the Safety Officer and specialists personally, although information and comments must be sought from those with detailed knowledge of each working area, such as local managers and members of the Safety Committee. The Safety Officer must pull the information together, and try to ensure that the amount of attention given to each type of hazard is in proportion to the real risk. Known health and safety problems must be taken into account (see previous section 'Responding to problems'), but no hazard should be overlooked simply because no one has yet been injured or taken ill.

In some cases, the overview will directly identify actions to be taken. More commonly, it will conclude that a more detailed assessment by local managers is required to comply with the requirements of specific legislation such as the Control of Substances Hazardous to Health Regulations (COSHH). The product of the overview should be a written summary from the Safety Officer or specialist which identifies all the hazards present, and whether the associated risks have been adequately investigated and eliminated or controlled. Required actions such as detailed assessments or improvements to control measures should be clearly identified. Copies of the written summary should be provided to all interested parties such as local managers and members of the Safety

Table 1 Hazards likely to be present in the animal house and applicable legislation

Hazard	Some possible sources	Applicable legislation
Chemicals	Novel substances, anaesthetic agents, veterinary drugs, fixatives, disinfectants, fumigants, cage washing chemicals and other cleaning agents	The Control of Substances Hazardous to Health Regulations 1999 (COSHH) (Statutory Instruments 1999). COSHH regulations implement the European Carcinogens Directive (Council Directives 1990b)
Allergens	Rodents	The Control of Substances Hazardous to Health Regulations 1999 (COSHH)
Infections	Zoonotic infections, work involving living organisms, e.g. vaccine development, testing of biopesticides, genetic modification	The Control of Substances Hazardous to Health Regulations 1999 (COSHH). COSHH regulations implement the European Biological Agents Directive (Council Directives 1990c). The Genetically Modified Organisms (Contained Use) Regulations 1992 (Statutory Instruments 1992c) implementing the European Contained Use of Genetically Modified Organisms Directive (Council Directives 1990d)
Radiation	X-ray equipment, radiochemicals	Radioactive Substances Act 1993. The Ionising Radiations Regulations 1985 (Statutory Instruments 1985)
Electrical hazards (shock, burns)	Faulty or unsuitable electrical equipment. Water in contact with electrical equipment	The Electricity at Work Regulations 1989 (Statutory Instruments 1989a)
Physical injury	Animal bites, needlestick injuries, cuts from sharp objects, falls, burns	The Management of Health and Safety at Work Regulations 1992 (Statutory Instruments 1992a). May expose the body to allergenic, chemical or infectious hazards covered by COSHH
Manual handling hazards	Lifting and carrying dogs, primates, cages, batteries, animal diet and bedding	Manual Handling Operations Regulations 1992 (Statutory Instruments 1992b) implementing the European Manual Handling Directive (Council Directives 1990a)
Noise	Dogs barking	The Noise at Work Regulations 1989 (Statutory Instruments 1989b) implementing the European Noise at Work Directive (Council Directives 1986)
Fire	Faulty electrical wiring or equipment, flammable liquids	Fire Precautions Act 1971

Committee. Senior management should be briefed by the Safety Officer or the safety champion since there are almost certain to be at least some resource or policy implications. A well-written summary will provide the basis for a shared understanding of the organization's safety problems and priorities, and the subsequent programme of action.

COSHH

COSHH will probably account for the greatest number of detailed assessments required. Anyone intending to perform a COSHH assessment must be competent to do so and must be guided by the latest edition of the Approved Codes of Practice: at the time of publication this is the 1999 edition (Health and Safety Commission 1999), published to accompany the 1999 revision of the COSHH Regulations (Statutory Instruments 1999).

COSHH Regulations implement the European Carcinogens Directive (Council Directives 1990b). In consequence, COSHH adopts a more prescriptive stance to carcinogens than to other hazardous substances and lists eight specific and mandatory control measures (see pp. 169–70).

Almost inevitably, the first stage of any COSHH assessment is to assemble a list of the hazardous substances present in an area, and the principal hazards presented by each. For a chemical or chemical product, the starting points for information gathering are the container label, the supplier's Safety Data Sheet and the current edition of Guidance Note EH40, Occupational Exposure Limits (Health and Safety Executive, annual publication). For many chemicals, these will provide all the information that is needed. For others, it may be necessary to search the scientific literature. Some useful places to look are the *Handbook of Occupational Hygiene* (Harvey 1993), *Dictionary of Substances and their Effects* (Richardson 1992–95), *The Sigma-Aldrich Library of Chemical Safety Data* (Lenga 1987), and, for pharmaceuticals, *Martindale* (Reynolds 1993). For novel substances, supplied by a sponsor or another department for testing, it is useful to have a standard questionnaire form requesting available information on chemical structure, storage conditions, physical properties, toxicology and any reactions observed in humans. This form should be completed by the supplier of the substance, so that a COSHH assessment can be performed before work begins.

The above information can be collected by the local manager of the area where the substance is to be used, or an appropriate member of the Safety Committee, but it is advisable for the Safety

Officer to keep a central record. This will help to ensure consistency between work areas, and prevent unnecessary duplication of effort in data gathering. It will probably be convenient to store the information in a computerized database: in larger establishments, it may be worthwhile making it available to multiple users via a network.

It must not be forgotten that a substance does not have to be pure or have a defined chemical structure in order to fall within the scope of the COSHH Regulations. Thus animal allergens and zoonotic infections are covered by the regulations, and the risks associated with these must be assessed. Two publications from the ESAC (HSC Education Services Advisory Committee 1990, 1992) give useful information.

The second stage of COSHH assessment is to consider how exposure to each hazardous substance may occur. Potential routes of exposure need to be considered: depending on the nature of the substance and the conditions of use, exposure may occur by inhalation, ingestion, eye or skin contact or through cuts or needlestick injuries. The groups of people who may be exposed must be identified. The potential for exposure may not be limited to individuals performing the work, but may extend to other groups such as maintenance engineers and laundry staff.

The third stage of COSHH assessment is to identify the methods of preventing or controlling exposure which are available within the area, and examine the existing or proposed work practices which should have been documented already as described under 'basic documentation'. This is a good opportunity to check that SOPs and other documents are up to date, i.e. that no changes have occurred since they were last revised. Control measures which may be provided in an animal house include:

- a high standard of room ventilation, with no recirculation of extracted air;
- filter top cages;
- negative pressure safety cabinets (air is removed through an exhaust duct, creating an inward air flow at the working face);
- total containment in isolators;
- downdraught ventilated necropsy tables to remove fixative fumes;
- carbon filters on anaesthetic equipment to prevent escape of volatile anaesthetics into the room atmosphere;
- protective clothing, including worksuits, gowns, gloves, respirators and eye protection.

The three stages of COSHH assessment described so far are relatively easy, although some time may need to be spent gathering the required information. The next stage is much more difficult. It consists of taking the information on the hazards of each substance, the potential routes of exposure, the groups of people who may be exposed and the actual or proposed safety precautions, and deciding whether control of exposure is adequate. In general, an adequate level of control is one that would prevent any adverse health effect in nearly all of the population, even if the work is performed repeatedly. If control of exposure is not adequate, it must be improved, or the work cannot be allowed to proceed. This decision cannot be avoided or fudged. In all except the most straightforward and obvious cases, someone is going to have to sign a piece of paper saying one of the following things:

- that the safety precautions are sufficient, and the work can go ahead;
- that the safety precautions appear to be sufficient; the work can go ahead, but further investigations such as exposure monitoring or health surveillance are required;
- that the work can go ahead subject to specified improvements in the safety precautions;
- that the work cannot go ahead.

How is the decision to be made? The decision frequently requires judgement, based on experience. The Safety Officer must either make the decision, or, perhaps more likely, endorse that made by the local manager. Fortunately, various criteria and items of guidance are available to help in the decision-making process:

- for a substance assigned an occupational exposure standard (OES) in Guidance Note EH40, control of exposure by the inhalation route can be regarded as adequate if exposure does not exceed the OES.
- for a substance assigned a maximum exposure limit (MEL) in EH40, control of exposure by the inhalation route can only be regarded as adequate if the level of exposure has been reduced so far as is reasonably practicable and in any case to below the MEL.
- for a carcinogen, the control of exposure cannot be regarded as adequate unless all of the following eight control measures have been implemented:

1) the total enclosure of the process and handling systems unless this is not reasonably practicable;
2) plant, processes and systems of work which minimize the generation of, or suppress and contain, spills, leaks, dust, fumes and vapours of carcinogens;
3) limitation of the quantities of a carcinogen at the place of work;
4) keeping the number of persons who might be exposed to a carcinogen to a minimum;
5) prohibiting eating, drinking and smoking in areas that may be contaminated by carcinogens;
6) the provision of hygiene measures including adequate washing facilities and regular cleaning of walls and surfaces;
7) the designation of those areas and installations which may be contaminated by carcinogens, and the use of suitable and sufficient warning signs;
8) the safe storage, handling and disposal of carcinogens and use of closed and clearly labelled containers.

- for animal allergens, control of exposure can be regarded as adequate if good practice, as described in Chapter 2 of this book and by the ESAC (HSC Education Services Advisory Committee 1990) has been followed, **and** the results of health surveillance indicate a low incidence of laboratory animal allergy.
- for infectious hazards, control of exposure can be regarded as adequate if good practice has been followed, **and** there has been no history of infection or illness among staff suggestive of inadequate containment. For information on good practice, refer to Chapter 3 of this book, to the COSHH Approved Codes of Practice and to the advice given by the ESAC (HSC Education Services Advisory Committee 1992). For primate zoonotic infections refer to an MRC booklet (Medical Research Council 1990) updated recently by ACDP (1998). For recommended containment measures when working with pathogens, refer to ACDP guidance (HSC Advisory Committee on Dangerous Pathogens 1995, ACDP 1997).

How does one approach the assessment of work with a chemical substance that does not have an established OES or MEL? For a commercially available substance, the supplier's Safety Data Sheet will specify handling precautions, exposure controls and personal protection measures: these should be observed. For a novel substance, it may be appropriate to set an in-house standard by reference to substances of similar toxicity to animals which do have OESs or MELs. If there is any doubt about the acceptability of the levels of airborne contaminant remaining after other methods of exposure control have been employed, operators will usually be required to wear respiratory protection against dust or vapours, as appropriate.

How does one approach the assessment of work with a chemical substance where the greatest risk is by a route other than inhalation? It is essential to take into account the consequences to the individual of exposure to the substance. Thus, if the substance is corrosive, and not completely contained, the greatest risk is to the eye where a splash may cause permanent damage. The highest standard of eye protection available is provided by chemical safety goggles, so these must be worn. If the substance is irritant or sensitizing to the skin, or toxic by absorption through the skin, all areas of skin likely to come into contact with the substance must be protected. Gloves must be of a material which is of low permeability to the substance. Aromatic and chlorinated solvents can permeate rapidly through rubber and vinyl gloves. For these solvents, it is essential to specify gloves made of nitrile rubber, polyvinyl alcohol (PVA) or aluminium foil laminate.

Generally, a separate COSHH assessment is made for each hazardous substance present in an area. However, this process can become cumbersome where large numbers of substances are received for testing, as in pharmacology or toxicology departments. A workable approach in this case is to assign each compound to one of several hazard classes based on its known or expected hazardous properties, and to check by assessment that the written safety instructions for each hazard class are appropriate.

Further advice on performing COSHH assessments may be obtained from an HSE booklet (Health and Safety Executive 1993).

Radiation

When dealing with radioactive hazards, the first step will be to document equipment which emits radiation, and radiochemicals which are present. In order to use radioactive substances, the organization must possess a Certificate of Registration issued by Her Majesty's Inspectorate of Pollution (HMIP) under the Radioactive Substances Act 1993. This certificate will specify the identity and maximum amount of each radionuclide that may be held. Detailed records must be kept of each receipt and disposal of a radioactive substance; records of current stocks must also be kept for comparison with the holding limits given in the certificate.

The Ionising Radiations Regulations (Statutory Instruments 1985) give dose limits for ionizing radiation which must not be exceeded, and require every employer to appoint one or more employees as radiation protection supervisors to ensure that work is carried out in accordance with the regulations. The radiation protection supervisor will assess the likely degree of exposure of employees to radiation and, if this exceeds thresholds given in the regulations, will recommend to the employer that supervised or controlled areas be established. In any case where an employee is exposed to a dose rate, averaged over one minute, which exceeds 7.5 μSvh^{-1}, or the employer has designated a controlled area which persons enter, a radiation protection adviser must be appointed to give specialist advice.

Electrical hazards

For control of electrical hazards, it is essential that all electrical systems should be designed, installed and maintained to the standards required by the Electricity at Work Regulations (Statutory Instruments 1989a). Installation should be carried out to the recommendations of the IEE Wiring Regulations (Institution of Electrical Engineers 1991). The maintenance department will generally be responsible for maintenance and testing of electrical systems, including portable electrical appliances, at appropriate intervals. The Safety Officer should make sure that suitable arrangements are in place. Correct behaviour on the part of users is also essential if the creation of electrical hazards is to be avoided. Fitting of plugs, alterations to wiring and repair of defective electrical equipment must only be undertaken by suitably competent people. Electrical equipment that is not specifically designed to be waterproof must not be allowed to become wet. Any equipment that is damaged or defective must be taken out of service and arrangements made for its repair. Particular care should be taken to avoid damage to flexible cables. Laboratory rodents are quite capable of chewing through PVC insulation. Flexible cables can also be damaged by cage batteries, trolleys and sack barrows. Assessment of electrical hazards is largely a matter of checking that appropriate precautions are being followed.

Physical injury

Assessment of the risks of physical injury in the animal house must take into account that it is often not the injury itself that is the main risk to health, but damage to skin, for example, will enable allergens, hazardous chemicals, infectious agents or radioactive substances to enter the body. The most serious risks are likely to be from non-human primate zoonotic or experimental infections. Arrangements must be made for first aid and medical treatment of injury or illness at work, including animal bites and their effects, and the results of microbiological screens performed on the animals should be readily available. Consider excluding from work with non-human primates that may be carrying potentially serious zoonoses any staff whose immune systems are compromised by illness or medical treatment, or who are pregnant. Rodent bites do not generally result in significant harm, but can trigger a serious asthma attack in someone who is sensitized and already has asthma. Consider excluding individuals with a history of asthma from work with rodents. Accidents that introduce clinically significant amounts of hazardous chemicals or radioactive substances into the body are relatively unlikely.

The frequency of animal bites should be minimized by thorough training in animal handling techniques. Suitable gloves should be worn that neither decrease dexterity nor disturb the animals. To reduce the risk of needlestick injuries, hypodermic needles should not be recapped after use, but should be placed immediately in a suitable 'sharps' container. As a general policy, a sharp object should never be used where a blunt one would serve equally well. For example, do not use a hypodermic needle to transfer liquids from one open container to another: a blunt-ended pipette or cannula will do just as well. Do not use scissors with sharp points if scissors with rounded ends would be equally effective. Assessment of the risks of physical injury should include a check that these measures are in place.

Other physical injury risks which need to be assessed include falls and burns, e.g. from autoclaves, cagewashers and hot articles and pipes.

Manual handling

For hazards that are the subject of other legislation, e.g. manual handling, the process of assessment is often similar to that which occurs under COSHH. The hazards present are identified, the existing or proposed methods of working are documented, and a decision is made as to whether the situation is satisfactory. Thus in the case of assessment of risks from manual handling tasks, it is necessary to list all the tasks that involve moving heavy or cumbersome items such as cages, other equipment or animals, and describe how these tasks are performed. Important aspects are the height through which an item has to be lifted, whether it can be held close to the body, whether stooping or twisting movements are required, and how often the task has to be performed. Various methods of improvement may be considered if the assessment is unfavourable, e.g. provision of mechanical lifting equipment, provision of a sack barrow or trolley, storage of materials at a more convenient height, provision of steps so that people do not have to stretch above head height, removal of obstructions so that the lifter can adopt the correct technique, and lifting shared with a second person. Detailed guidance on performing manual handling assessments has been published by HSE (Health and Safety Executive 1992). This includes numerical guidelines for weights that may be lifted and lowered by nearly all men. For women, the guideline figures should be reduced by one-third.

Noise

The required actions to assess noise and reduce exposure are set out in the Noise at Work Regulations (Statutory Instruments 1989b). Noise hazards may come from animals, particularly barking dogs, from heating and ventilation plant, and from movement of cage batteries, particularly in cage-washing areas with dimpled tile floors. The first step in assessment will be to recognize the significant sources of noise and the people who may be exposed to them. Personal noise dosimetry will then be required to determine daily noise-exposure levels. Since the dosimetry will need to be repeated only if noise levels change, it will often be more convenient to have these measurements carried out by an outside consultant rather than acquire the equipment and expertise necessary to perform them and interpret the results. If the daily noise exposure exceeds 85 dB(A), 'the first action level', employees must be provided on request with personal ear protectors. If the daily exposure exceeds 90 dB(A), 'the second action level', or the peak sound pressure exceeds 200 pascals, 'the peak action level', the area must be designated an ear-protection zone, with appropriate safety signs, and anyone entering it must wear personal ear protectors. The noise of barking beagles is very likely to exceed the second action level, so any area where beagles are housed will probably be designated an ear-protection zone. The regulations require that noises louder than the second or peak action level must be reduced at source, so far as is reasonably practicable, but it is not considered reasonably practicable to reduce at source the noise made by a barking dog!

Fire

Fire is treated differently from the other hazards mentioned above. An application for permission under Building Regulations to construct or modify a building will be referred to the fire authority for their requirements. Later, the Fire Prevention Officer will examine the completed building. If more than 20 people work in the building, or more than 10 people work elsewhere than on the ground floor, a fire certificate will be required (Fire Precautions Act 1971). In effect, the fire authority performs the risk assessment and specifies the safety measures required. The role of the user of the building is limited to providing information, applying for a fire certificate if one is required, and complying with the fire authority's instructions. These will usually relate to fire extinguishers, fire alarm systems, escape routes, signs, evacuation procedures and fire drills.

Where flammable gases and vapours are present, equipment will need to be spark-free. Examples are refrigerators used to store carcasses of animals killed by overdose of ether. (NB. This method of euthanasia is not recommended, for animal welfare and safety reasons.)

Risk reduction

For all hazards, if the assessment shows that the risk needs to be reduced before the work can go ahead, the measures necessary to achieve this reduction must be clearly identified. Vague statements of general intent will not do. The measures required may be of several different types:

- use of safer alternative substances or processes;
- improved containment;
- changes to working practices (accompanied, of course, by changes in the relevant written instructions);
- improved personal protective equipment;
- vaccination of staff, e.g. against tetanus, rabies or tuberculosis;
- further training;
- exclusion of certain categories of staff, e.g. pregnant women, asthmatics, laboratory animal allergy sufferers.

Records

Adequate records of all assessments need to be kept. As already explained, the assessment process results in a decision. Whoever makes the decision must be prepared to explain and defend it at any time, in court if necessary. It follows, therefore, that sufficiently comprehensive records must be kept to enable the then current state of knowledge and the process of reasoning which led to the decision to be reconstructed accurately at any time in the future. Once again using COSHH as an example, the records would need to cover, for each substance present in an area:

- the name of the substance;
- the available information on its properties;
- the work to be performed;
- the containment measures to be used;
- the personal protective equipment to be used;
- any categories of staff that are to be excluded;
- available data on exposure levels;
- available data from health surveillance;
- the conclusion of the assessment, i.e. the decision reached;
- the process of reasoning by which the conclusion was derived (if not obvious);
- the signature of the person who performed the assessment, and the date.

It is easy to generate pages of records for each hazardous substance, and some people have done exactly that. The pragmatic Safety Officer will realize that there are several legitimate short-cuts that will help to minimize the quantity of paper produced. Thus, if there are two or three standard sets of safety precautions (containment measures and personal protective equipment) in use in an area, and each is applicable to a number of hazardous substances, it is better to describe these sets of safety precautions in an SOP and just cross-refer to the appropriate one in the assessment. If standard rules, documented in an SOP, are used to link given hazardous properties with given conclusions about control measures, much narrative discussion can be avoided. The following are examples of such standard rules:

- If the substance is corrosive, chemical safety goggles will always be required and appropriate protective clothing should be worn to protect the skin.
- If the substance is a volatile liquid without an established OES or MEL, a vapour respirator will always be required.
- If the substance is mutagenic or teratogenic, pregnant women will always be excluded.

All records should be retained for at least 40 years.

Whenever a new hazard is to be introduced into an area, or the work being done is to be changed sufficiently to alter the risk from an existing hazard, an assessment must be performed before the new work starts. Thus, in the case of hazardous substances, a COSHH assessment will be required for each new substance, and also whenever the nature and scale of the work with an existing substance changes sufficiently to make the existing assessment no longer valid. The Approved Codes of Practice (Health and Safety Commission 1999) state that COSHH assessments should be reviewed at least every five years.

Implementing required changes

Some changes have minimal resource implications. Examples include the use of containment measures or personal protective clothing which are already available, or a change in the brand of eye or respiratory protection issued to staff. Other changes may be accommodated conveniently within existing budgets, e.g. the provision of videos and other training aids. Sometimes, however, an assessment will reveal the need for expensive and previously unforeseen changes. Capital expenditure may be required to install extra containment devices and to improve ventilation. Changes in working practices may be required which reduce the productivity of employees and hence increase running costs; for example, placing animals in an isolator will dramatically slow all husbandry and dosing tasks. The need for expensive changes will rarely be welcome news to the management of the organization. The Safety Officer, aided and abetted by the safety champion – if there is one – will need to be at his or her most persuasive. In particular, it will be necessary to establish three points clearly:

- why it is necessary to make a change, e.g. because the present situation contravenes one or more regulations or has attracted adverse comment from HSE, health problems have occurred, new

information on the hazardous properties of a substance has been received, or it is necessary to work with a new substance more hazardous than any currently in use;
- that the proposed change will achieve the desired effect;
- that the proposed change is the most cost-effective response to the problem.

Expert advice from a consultant may be required to tackle one or more of these points.

The ability to solve difficult problems inexpensively is a skill that is (or at least should be) valued highly by any organization with limited resources. The following examples may serve to illustrate the point:

- For any open-fronted containment device (safety cabinet, fume hood, etc.) the degree of containment will be related to the velocity of the air at the face opening – the 'face velocity'. This should normally be between 0.5 and 1.0 metres per second (0.7 to 1.0 ms^{-1} for a British Standard BS 5726 Class I microbiological safety cabinet). Hopefully, all cabinets now used for work with pathogenic microorganisms comply with BS 5726, but a variety of other open-fronted hoods is used for chemical hazards. Many of these hoods may have face velocities too low for satisfactory containment. What is to be done? If there is spare capacity in the air extraction and supply systems to the area, the volume of air extracted per minute can be increased. In reality, this is often not the case: either the designer of the building did not provide any spare capacity, or the occupants have used it up by putting in more hoods than the original design envisaged. To increase the amount of air supplied to and extracted from a building will involve major expenditure on new plant. At this stage, the Safety Officer should ask serious questions about the face openings of all the hoods present in the area: if these can be reduced without making the hoods difficult to use, the face velocity can be increased without extracting any more air. In some cases, an unsatisfactory hood can be made to perform to an acceptable standard simply by mounting an extra perspex panel on the front to reduce the face opening.
- Some airborne chemical hazards arise from point sources. Examples are anaesthetic equipment, and decanting of waste chemicals into drums for disposal. For point sources, local exhaust ventilation positioned as close as possible to the source will provide the best control for the smallest volume of extracted air (and hence lowest cost).

Before implementing or seeking authorization to implement any change, the local manager or Safety Officer must check its practicability. This will inevitably involve close consultation with all levels of management and staff of the area concerned and appropriate members of the Safety Committee. If possible, a trial of the new safety precautions should be carried out before a final decision is made. If a change in personal protective equipment is being considered, it should always be possible to obtain enough samples of the new type for a trial. It will also be necessary to consider the transition period when changes are being made. If building works or installation of new fixed equipment are required, how long will these take? How much noise, dust, interruption of services (electricity, hot and cold water, heating and ventilation, etc.) and interruption of ongoing work will there be? Will these be acceptable when animals are present in the building?

Where possible, changes to documentation such as SOPs should be prepared before the actual change in safety precautions is made, but urgently needed safety improvements should not be delayed while this is done. If necessary, make the change straight away and catch up with the paperwork as soon as possible afterwards.

All staff affected by the change should be trained in the new method, and the reasons for the change should be explained to them. Local managers will need to check periodically that there has been no sliding back into the old ways. Sometimes there is resistance to change, usually along the lines that the new precautions make the work too slow, or are too uncomfortable. Consultation and trials in advance of making the change will help to prevent large-scale resistance occurring. Even so, there will sometimes be objections from isolated individuals. It is important to establish whether these objections are justified, since individual objections to particular brands of personal protective equipment are sometimes valid. For example, some manufacturers of eye protection describe their one size of product as 'universal fit', but this may be overly optimistic. If a respirator is available in one size only, it may not give an adequate fit for people with differing shapes and sizes of face. A responsible employer will try hard to meet the legitimate requirements of individual staff, even if it means stocking more than one size or brand. A small minority of staff may be allergic to the starch used as a lubricant in disposable gloves. It is possible to obtain lubricant-free gloves or the problem may be tackled by supplying cotton liners to be worn inside the gloves.

Having dealt with the justified objections to new safety precautions, what should be done about unjustified ones? Where appropriate, further instruction and training should be given.

Eventually, it may be necessary to invoke disciplinary procedures, with their ultimate sanction of dismissal for persistent offenders, but this should be regarded as a last resort. Nevertheless, the organization must protect its own position by insisting that satisfactory safety precautions are taken.

Sufficient records must be kept to ensure that it is possible, in the future, to determine what safety precautions were being employed in any given area, for any given task, on any given date. To achieve this, all safety SOPs or other written safety instructions should have a commencement date, and an archive file of all superseded versions should be kept.

Maintenance and testing of control measures

All equipment and plant used to eliminate or control exposure to hazards must be in good working order if they are to deliver the expected standards of protection. Engineering measures such as general room ventilation, safety cabinets, fume hoods, isolators and ventilated necropsy tables must be tested to ensure that they are functioning correctly. The required tests may monitor air flow rates, integrity of filters, pressure drop across filters, and effectiveness of containment. Such tests will be performed after installation, as part of the commissioning process, then regularly at intervals of six or 12 months. Where appropriate, devices will be fitted to give an immediate alarm to the user in the event of air-flow failure or filter blockage. In addition to routine testing, there will often be a planned programme for replacement of items such as filters which have a limited life.

The COSHH Regulations also brought in a requirement for visual examination of respiratory protective equipment (RPE) at least once every month. In the case of RPE incorporating compressed gases or electric motors or fed by air line, tests are required to ensure that the equipment is functioning correctly. In many cases, users of RPE have avoided these requirements by switching to disposable respirators where these offer sufficient protection.

The Safety Officer will wish to ensure that there are written procedures specifying:

- the control measures that require regular testing and/or maintenance;
- the tests and maintenance actions required, and the frequency with which they are to be carried out;
- who is to perform the various actions;

- how the results of the tests, and completion of the maintenance actions, are to be documented.

Records of tests must identify the equipment or system concerned, and show who did what and when, and with what result. They must document any faults that were found. They must show whether the equipment or system was still operating at the required performance or, if not, the repairs required to achieve that performance. Such records must be kept for at least five years. The Safety Officer may wish to audit such records from time to time to ensure that the equipment is in fact being tested and maintained at the specified intervals, that the records are being kept, and that any required remedial action is being taken.

Exposure monitoring

It may be necessary to measure the extent of exposure of staff to one or more hazardous agents, e.g. dusts, vapours, noise or radiation. Such monitoring may be required as part of the initial risk assessment process, or to demonstrate continuing compliance with exposure limits. It need not be performed if there is no suitable method readily available, as in the case of animal allergens. It need not be performed if you can be sure that your safety precautions are adequate without doing it. For chemical hazards, it may be necessary to perform exposure monitoring for only a limited number of substances in order to demonstrate the adequacy of control measures. Thus, for organic solvents, one might choose to monitor a few substances characterized by at least two of the following:

- low boiling point;
- low MEL or OES;
- used in large quantities.

In any area in which a formaldehyde-based fixative is used, it will almost always be necessary to monitor formaldehyde levels in the air at regular intervals as these fixatives tend to be used in large quantities, and the MEL for formaldehyde is very low at 2.0 ppm. Furthermore, this limit is applied to short-term exposure, averaged over 15 minutes, as well as to long-term exposure, averaged over eight hours. Formaldehyde levels should also be monitored after fumigation.

For hazardous dusts, one might choose a few substances which are fine powders, and have a low MEL or OES or are used in large quantities. These selected substances provide a worst-case test of the available control measures.

Where the assessment shows that monitoring of exposure to chemicals is required, it should be carried out at least every 12 months. Written procedures should document the measuring and sampling methods to be used, the location and frequency of sampling, and how the results are to be interpreted. Samples for monitoring of airborne contaminants are best taken from the breathing zone of the person performing the task being studied. The measurement may determine the concentration of contaminants at a single point in time, or may determine the cumulative exposure over a period of time, up to eight hours. A cumulative exposure result is usually divided by the duration of exposure to give a time-weighted average concentration. Single-point measurements of vapours may be made by a suitable sensor and meter, or by a detector tube, through which air is drawn by bellows or a gas syringe. Cumulative measurements of vapours may be made by trapping the vapour in a suitable medium for analysis later, or by means of an indicator badge that undergoes a cumulative colour change. Dust samples for cumulative measurements are collected by means of a pump which draws air through a filter at an accurate, predetermined flow rate. The material collected on the filter is weighed or subjected to chemical analysis later.

Where the organization employs an occupational hygienist, he or she will be responsible for exposure monitoring. If no such person is employed, the Safety Officer will probably find it useful to designate one individual to be responsible for the monitoring equipment, including training others in its use, when required, and maintaining stocks of consumables such as indicator tubes and dust-collection filters.

Records of exposure monitoring should identify the substance monitored; the date and time, or time period, that each sample was taken; the methods used for sampling and analysis; the area in which the sample was taken; the name of the person from whose breathing zone the sample was taken; the task that was in progress at the time; and the result obtained. These records should be kept for at least 40 years.

Since MELs and OESs are expressed in terms of time-weighted averages, cumulative measurements of exposure are easily interpreted. Calculating a time-weighted average concentration from a series of single-point measurements is prone to error, and probably best not attempted. Nevertheless, single-point measurements can be of value if they are selected to include the times when levels are likely to be at their peak. If the peak exposure concentration is below the numerical value of the exposure limit, you are certainly in compliance, since the time-weighted average concentration can only be lower. On the other hand, if the peak concentration is above the exposure limit, it is necessary to proceed to a cumulative measurement, which will give you a time-weighted average result for direct comparison with the limit.

Monitoring for noise has already been described under 'Organizing risk assessment'. Monitoring for exposure to radioactivity is often performed by means of film badges, which are worn at all times that a person is in an area where a radiation hazard is present.

All results of exposure monitoring should be made available to the relevant staff and to the Safety Committee.

Health surveillance and health records

Health surveillance of employees is required where exposure to a hazardous substance is reasonably likely to give rise to an identifiable disease or adverse health effect that can be detected by available techniques. Since laboratory animal allergens are reasonably likely to give rise to the detectable disease of laboratory animal allergy (LAA), all animal house staff and those using animals require health surveillance. The basic minimum level of health surveillance consists of a questionnaire, completed by each member of staff, which asks about a range of symptoms that may be indicative of LAA: breathing difficulties, wheezing, skin rashes, sore and watery eyes, sneezing and running nose. This questionnaire should be completed at least annually. Staff should also be encouraged to report such symptoms to their supervisor if and when they occur. The occurrence of such symptoms should trigger medical follow-up, which may include measurements of lung function such as the peak flow rate during forced expiration: this is measured by taking a deep breath and blowing into a peak flow meter, a small portable instrument. Although COSHH requires health surveillance of employees only, the ESAC recommends that it should be extended voluntarily to students and researchers exposed to animal allergens, e.g. in higher education (HSC Education Services Advisory Committee 1990). The Safety Officer should consider whether health surveillance should be extended to other staff such as maintenance engineers and laundry workers who may also be exposed to animal allergens.

Clinical data from health surveillance is confidential to the doctor and the individual concerned, but the employer must be informed of the conclusion, in terms of the employee's fitness for work. In mild cases of LAA, it may be acceptable for the individual to continue working in the animal house, preferably with upgraded safety precautions; but in more serious cases, particularly those involving breathing difficulties, it will be necessary for the individual to give up such work immediately in order to avoid permanent damage to health. An employee may be reluctant to report symptoms if this is likely to lead to loss of his or her job. This is less likely to be a problem if the organization has a record of sympathetic treatment of previous cases, e.g. moving affected people to other jobs where they will not be exposed to animal allergens. It should also be made abundantly clear to staff during training that ignoring symptoms can permanently damage health, and is simply not worthwhile.

Biological monitoring, i.e. the measurement of hazardous chemicals or their metabolites in body fluids, excreta or expired air, may be used to determine the level of exposure to chemicals. This is particularly useful where absorption can occur by routes other than inhalation, i.e. where monitoring of airborne contaminants will give an incomplete picture of total exposure. Measurement of the relevant radionuclide in blood and urine samples is useful if it is suspected that someone has absorbed a radioactive substance as a result of dust inhalation or a needlestick accident.

Employees must be provided with a summary of the collective results of health surveillance, in a form which preserves the anonymity of the individuals concerned. Questionnaire data can be summarized in terms of the incidence of positive answers to each question. It is easier to interpret the results if one has access to similar data from a control group of staff working in another part of the organization, and who are not exposed to animal allergens. The statistical significance of the difference in incidence of a symptom between the animal house staff and the control group can then be determined by Fisher's exact test (Conover 1980).

For all employees undergoing routine health surveillance, and those who have worked with known or suspected carcinogens, the employer is required to keep health records. The Safety Officer may persuade the personnel department to perform this task, since they will already have most of the information. The information to be kept for each individual is as follows:

- surname, forenames, sex, date of birth, permanent address, post code, National Insurance Number, date of commencement of present employment;
- a historical record of jobs requiring health surveillance in this employment;

- conclusions of health surveillance procedures, who carried them out and when.

The health record must be kept for at least 40 years from the date of the last entry.

Animal house management should also have data on the health status of its animals. Of particular relevance to human health are the results of screens for zoonotic infections in primates. For example, if a person is bitten by a primate, the *Herpesvirus simiae* (B virus) antibody titre of the animal, a measure of infection by the virus, may be the deciding factor in the decision to initiate precautionary treatment with an antiviral drug.

Safe disposal of waste

Proper disposal of waste is essential to protect the health and safety of staff, waste disposal contractors and the general public, and to prevent damage to the environment. The legal framework for disposal of waste is now provided by the Environmental Protection Act 1990. This Act imposes a duty of care on everyone who produces, imports, carries, keeps, treats or disposes of commercial or industrial waste. The duty of care also applies to everyone involved with household waste, except the householder who produced it. Those subject to the duty must try to prevent any other person disposing, treating or storing the waste illegally, must contain the waste, must transfer it only to an authorized person, and must provide an appropriate written description in the form of a transfer note. Detailed guidance is given in a code of practice (Department of the Environment *et al.* 1996).

To enable each type of waste to be disposed of by the best route, careful segregation at source is essential. In consequence, this aspect of waste disposal becomes part of the job of everyone who generates waste.

The following types of waste are likely to arise in an animal house:

- *waste from quarantined animals* – all must be destroyed by incineration.
- *clinical waste from non-quarantined animals* which requires disposal by incineration, e.g. carcasses, surplus veterinary drugs, novel substances supplied for testing and mixtures of these with liquid vehicles or animal diets, radioactive waste containing carbon-14 or tritium, sharps and syringes.

- *low-grade clinical waste for which incineration is not essential,* e.g. soiled bedding from non-quarantined animals. This can be sent to landfill.
- *aqueous liquid waste* such as surplus or used histological fixative solutions. Some Waste Regulation Authorities may permit discharge of this waste to sewer: this should always be confirmed in advance. In some instances, recovery of formaldehyde from used fixative solutions by distillation may be practicable. Alternatively, the waste may be removed from site by a contractor.
- *organic liquid waste* such as surplus volatile anaesthetics, which should be destroyed by incineration, either on-site or by a contractor. If the incinerator is not equipped with facilities for liquid feed, the liquid waste can be absorbed on a suitable solid carrier such as sawdust or cat litter. If a contractor is to remove the waste for off-site incineration, it is usually more economic to segregate halogenated waste, i.e. waste containing compounds of fluorine, chlorine, bromine or iodine, from other organic waste since halogenated waste attracts the highest charge for incineration.
- *non-hazardous general waste,* likely to be mainly paper.

The number of categories is not absolute, and will need to be varied to take into account local needs. The provision of adequate packaging is essential for segregation at source to work effectively. Polythene bags of different colours should be used to distinguish between solid waste which must be incinerated, and that which can be sent to landfill. Quarantine waste must be identified clearly. Suitable containers must be provided for collection of used sharps. It is best to collect aqueous and organic liquid wastes in pre-labelled drums. Care must be taken not to mix chemicals that may react together.

Arrangements for the collection and disposal of waste must be documented, and staff must be trained. The Safety Officer will wish to check periodically that the system is operating as intended.

Since 1 October 1995, clinical waste incinerators with a capacity of under one tonne per hour have been required to comply with new design and emission standards (Department of the Environment *et al.* 1995). An effect of these standards is a great increase in the capital cost of such incinerators. A predictable consequence is that off-site incineration of waste from laboratory animal facilities will become much more common than has been the case in the past.

Conclusion

This chapter has attempted to give an overview of the wide range of actions and responsibilities required of safety management, with a few tips on how certain problems can be approached. The Safety Officer will maintain a watching brief on changing legal requirements, collect information on hazards, perform or endorse assessments, respond to problems, propose policies, persuade senior management to provide finance for improvements (with the help of the safety champion, if there is one), be involved with the Safety Committee, and work alongside managers to ensure that suitable arrangements are in place for documentation, staff training, maintenance and testing of control measures, exposure monitoring, health surveillance and waste disposal. It is an important but sometimes difficult job.

References

Advisory Committee on Dangerous Pathogens (ACDP) (1997) *Working Safely with Research Animals: Management of Infection Risks*. Sudbury: HSE Books

Advisory Committee on Dangerous Pathogens (1998) *Working Safely with Simians: Management of Infection Risks*. Sudbury: HSE Books

Conover WJ (1980) *Practical non-parametric statistics*. 2nd edn. New York: Wiley, 167

Council Directives (1986) Council Directive of 12 May 1986 on the protection of workers from the risks related to exposure to noise at work (86/188/EEC) *Official Journal of the European Communities*, 24.5.86, No. L137, 28

Council Directives (1989) Council Directive of 12 June 1989 on the introduction of measures to encourage improvements in the safety and health of workers at work (89/391/EEC) *Official Journal of the European Communities*, 29.6.89, No. L183, 1–8

Council Directives (1990a) Council Directive of 29 May 1990 on the minimum health and safety requirements for the manual handling of loads where there is a risk particularly of back injury to workers (90/269/EEC) *Official Journal of the European Communities*, 21.6.90, No. L156, 9–13

Council Directives (1990b) Council Directive of 28 June 1990 on the protection of workers from the risks related to exposure to carcinogens at work (90/394/EEC) *Official Journal of the European Communities*, 26.7.90, No. L196, 1–7

Council Directives (1990c) Council Directive of 26 November 1990 on the protection of workers from risks related to exposure to biological agents at work (90/679/EEC) *Official Journal of the European Communities*, 31.12.90, No. L374, 1–12

Council Directives (1990d) Council Directive of 23 April 1990 on the contained use of genetically modified micro-organisms (90/219/EEC) *Official Journal of the European Communities*, 8.5.90, No. L117

Department of the Environment, Scottish Office and Welsh Office (1995) *Secretary of State's Guidance – Clinical waste incineration processes under 1 tonne an hour*. London: HMSO

Department of the Environment, Scottish Office and Welsh Office (1996) *Waste Management: The Duty of Care, A Code of Practice*. London: HMSO

Environmental Protection Act (1990) London: HMSO

Fire Precautions Act (1971) London: HMSO

Harvey B (ed-in-chief) (1993) *Handbook of occupational hygiene*. Kingston upon Thames, Surrey: Croner Publications

Health and Safety at Work etc. Act (1974) London: HMSO

Health and Safety Commission (1992) *Management of health and safety at work: Approved Code of Practice*. London: HMSO

Health and Safety Commission (1999) *Control of Substances Hazardous to Health Regulations 1999: Approved Codes of Practice* L5. Sudbury: HSE Books

Health and Safety Executive (1991) *Your firm's injury records and how to use them*. Bootle, Merseyside: HSE

Health and Safety Executive (1992) *Manual handling: guidance on regulations*. London: HMSO

Health and Safety Executive (1993) *A step-by-step guide to COSHH Assessment*. London: HMSO

Health and Safety Executive (1996) *A Guide to the Reporting of Injuries, Diseases and Dangerous Occurrences Regulations 1995 (RIDDOR)* (L73). London: HMSO

Health and Safety Executive (1997) *Successful health and safety management*. London: HMSO

Health and Safety Executive (1998) *Managing health and safety – Five steps to success*. Sudbury: HSE Books

Health and Safety Executive (annual publication). *EH40/(year): Occupational exposure limits (year)*. Sudbury: HSE Books

HSC Advisory Committee on Dangerous Pathogens (1995) *Categorisation of biological agents according to hazard and categories of containment*. 4th edn. Sudbury: HSE Books

HSC Education Services Advisory Committee (1985) *Safety policies in the education sector*. London: HMSO

HSC Education Services Advisory Committee (1990) *What you should know about allergy to laboratory animals*. London: HMSO

HSC Education Services Advisory Committee (1992) *Health and safety in animal facilities*. London: HMSO

Institution of Electrical Engineers (1991) *Regulations for electrical installations*. 16th edn. Hitchin, Herts: IEE

Lenga RE (ed) (1987) *The Sigma-Aldrich Library of Chemical Safety Data*. 2nd edn. Milwaukee, WI: Sigma-Aldrich

Medical Research Council (1990) *The management of simians in relation to infectious hazards to staff*. London: MRC

Radioactive Substances Act (1993) London: HMSO

Reynolds JEF (ed) (1993) *Martindale: The Extra Pharmacopoeia*. 30th edn. London: The Pharmaceutical Press

Richardson ML (ed) (1992–95) *Dictionary of Substances and their Effects (7 vols)*. Cambridge: Royal Society of Chemistry

Social Security Act (1975) London: HMSO

Statutory Instruments (1977) No. 500. *The Safety Representatives and Safety Committees Regulations*. London: HMSO

Statutory Instruments (1985) No. 1333. *The Ionising Radiations Regulations*. London: HMSO

Statutory Instruments (1989a) No. 635. *The Electricity at Work Regulations*. London: HMSO

Statutory Instruments (1989b) No. 1790. *The Noise at Work Regulations*. London: HMSO

Statutory Instruments (1992a) No. 2051. *The Management of Health and Safety at Work Regulations*. London: HMSO

Statutory Instruments (1992b) No. 2793. *The Manual Handling Operations Regulations*. London: HMSO

Statutory Instruments (1992c) No. 3217. *The Genetically Modified Organisms (Contained Use) Regulations*. London: HMSO

Statutory Instruments (1995) No. 3163. *The Reporting of Injuries, Diseases and Dangerous Occurrences Regulations*. London: HMSO

Statutory Instruments (1997) No. 654. *The Good Laboratory Practice Regulations*. London: HMSO

Statutory Instruments (1999) No. 437. *The Control of Substances Hazardous to Health Regulations*. London: The Stationery Office

8

Legal requirements

Kevin Dolan

Contents

Introduction

The aim of this chapter is to set out the legal requirements relating to the health and safety duties of those in charge of and working with laboratory animals in the United Kingdom. There are legal obligations to the animal unit workers, to visitors and to the general public. As well as legislation concerned directly with animals, that dealing with, for example, machinery, electricity, accidents and waste disposal will be included. Such regulations, though applicable to much wider areas than the animal facilities, can be of immediate relevance to the health, safety and welfare of the animal worker and are involved in the control of hazards in the animal unit.

The precise significance of the terms 'health', 'safety' and 'welfare' is difficult to find in the vast amount of literature on the subject. Legal sources are singularly bereft of clear definitions of these three crucial terms. Authoritative definitions of 'health' seem to appear only in negative forms, e.g. 'That is injurious to health which makes sick people worse'. 'Health', in a positive sense, appears to be understood in a legal setting as the capability to use all one's natural faculties and the absence of any debilitating or diseased condition. There have been numerous authorities in which the meaning of the word 'safe' has been discussed; for example, 'safe is the converse of dangerous – means safe for all contingencies that may reasonably be foreseen, unlikely as well as likely, possible as well as probable'. It follows that related words, e.g. 'safety', 'safely', etc., should be construed accordingly. Work is considered safe if no cause of injury is reasonably foreseeable. Definitions of 'welfare' appear to be limited, but the listing of welfare facilities in Tolley's *Health and Safety at Work* Handbook (for most recent edition see Dewis 1998) such as provision of drinking water, clothing, accommodation and facilities for rest and eating meals are germane.

The enforcement of the law, the role of the Health and Safety Inspectorate and the imposition of specific penalties will be referred to where necessary.

It needs to be stressed that this chapter can in no way deal fully with the plethora of safety law. It can indicate only the salient features and point towards more authoritative and comprehensive presentation of the legislation elsewhere. In situations where there is the possibility of litigation, professional expert advice must be sought.

Law in general

United Kingdom (UK)

Statutes

Statutes, the main source of English Law, are the Acts passed by the House of Commons and the House of Lords, duly signed by the Sovereign. Examples of such statutes are the Factories Act 1961 and the Offices, Shops and Railway Premises Act 1963. Both of these Acts are gradually being replaced by regulations under the Health and Safety at Work etc. Act 1974 (HSWA) by virtue of Section 15 (s15) of the HSWA. Some specific provisions concerning safety in animal facilities appear in the Codes of Practice and the Conditions attached to the Certificates of Designation associated with the Animals (Scientific Procedures) Act 1986 (ASPA), the UK law that regulates the use and provision of laboratory animals.

Acts of Parliament may be restricted in their territorial application; e.g. Part I of the HSWA does not apply to Northern Ireland nor does most of Part III apply to Scotland. In some cases separate but similar Acts are enacted for the different countries within the UK. The government department responsible for the administration of a particular Act may vary from area to area. The Scottish or the Welsh Office may be responsible for the control of regulations which in England are supervised by a specific ministry. In future, national elected bodies, through the process of devolution, may have specific legislative roles.

Administration of the HSWA is the concern of the Health and Safety Commission (HSC) together with the Health and Safety Executive (HSE). The Health and Safety (Enforcing Authority) Regulations 1998 spell out the division of responsibility, between HSE and Local Authorities, for the enforcement of the HSWA and its relevant statutory provisions.

Other relevant legislation dealt with by authorities apart from the HSE will be referred to where appropriate. Such authorities are, for example, the Department of the Environment which deals with pollution, the Ministry of Agriculture, Fisheries and Food (MAFF) which deals with diseases of animals and quarantine regulations, the Department of Health which has issued codes about dangerous pathogens and the Fire Authorities which are concerned with fire regulations.

Crown immunity is a long-standing restriction on the force of Acts of Parliament. Unless stated within an Act itself and apart from exceptions mentioned in the Crown Proceedings Act 1947, crown premises and government departments are exempt from sanctions imposed under such Acts as the HSWA. The HSWA does, however, apply where appropriate to crown employees and the Crown is bound by the spirit and substantive provisions of the Act. The National Health Service (Amendment) Act 1986 effectively removed crown immunity from National Health Service (NHS) hospitals.

Subordinate legislation

Many modern Acts are enabling Acts, forming a framework into which subsidiary legislation may be inserted, without the full process of parliamentary enactment, by government departments or other authorities by virtue of powers conferred upon them by Parliament. The extent of such delegated legislative power is indicated within the Act itself. The HSWA, for instance, lays down the general duties of employers as regards safety but leaves the details of subsequent regulations to the Secretary of State or other legislative authorities. Subordinate legislation appears in the form of Statutory Instruments (SI), which are documents by which Her Majesty in Council or a Minister of the Crown exercises their statutory power to issue legislation pertinent to a particular area of activity. Statutory Instruments appear in the form of Orders, Regulations or Rules. Orders usually impose specific legal obligations within the context of an Act. Regulations spell out in detail the general obligations contained in the framework of an Act, are usually compulsory and are duly enforceable through the courts. In certain instances, like the Safety Representatives and Safety Committees Regulations 1977, a Regulation merely lays down a floor of rights which may be claimed by those for whose benefit it has been enacted. Some Regulations make provisions for exemptions to be granted from their requirements. Rules contained within a Statutory Instrument have the same legal status as Regulations but are usually associated with methods of administration of court procedure.

Approved Codes of Practice (ACOP) specify in detail or in a more liberal style than practicable or desirable in Regulations, the precise technical and other requirements to be considered in the observance of the law. ACOPs explain what, in particular circumstances, would be considered to constitute satisfactory compliance with the requirements of a general obligation. Authorised and Approved Lists may be associated with ACOPs. A failure on the part of any person to observe any provision of an ACOP is not in itself an offence, but the failure may be accepted by a court in criminal proceedings as proof that the person has contravened the Regulation to which the provision relates. It may be possible, however, for the accused to satisfy the court that he or she has complied with the

Regulation by some other means. Contravention of any provision of a relevant ACOP will put a person at a marked disadvantage in any subsequent litigation. Non-governmental codes published by professional or industrial bodies, e.g. the British Standards Institution, are merely expert recommendations on good practice in specified areas of work (see below).

Guidance notes are not mandatory and they do not have the evidential significance of ACOPs. They are official publications based on a wealth of practical experience. They indicate the way in which the HSE Inspectorate is likely to expect employers to operate. Further updated literature of an advisory nature is published frequently on a variety of safety topics by the HSC and the HSE in the form of leaflets and reports.

British Standards are codes of practice and specifications published by the British Standards Institution presenting detailed criteria for safety of equipment and operations in the workplace. They are valid guidelines for the full observance of safety laws in practice. Relevant ones are listed at the end of this chapter. Gradually their place will be taken by harmonized technical standards – EN (Euro Normes) – introduced by the Joint European Standards Institution. European CEN (Comité Européen de Normilisation) Standards are arranged in types:

> *A-type standards* lay down fundamental principles of machine safety.
> *B-type standards* deal with features such as safety distances, noise levels, guards and controls.
> *C-type standards* lay down safety requirements concerning specific types of machinery.

Europe

The supreme law in every Member State of the European Union (EU) is the Treaty of Rome (1957). Apart from the basic principles of the Treaty there are Regulations, Decisions and Directives of Ministers or the European Commission. Regulations are directly applicable to Member States without further enactment. Decisions become operative at once but may require action by Parliament to make them effective. Directives, with which we are concerned here, are binding in principle but need to be implemented locally by the Member State. All national law is to be interpreted in line with pertinent Directives even if the Directive is subsequent to a national legal enactment.

In the field of safety legislation, European law adds little over and above what is already demanded by UK law. The main safety Directive, the 'Framework Directive' (89/391/EEC), has become part of our legislation as the Management of Health and Safety at Work Regulations 1992. From the 'Framework Directive' came five other Directives resulting in the 'six-pack' of Regulations. The legislative basis of the 'six-pack' is Article 118A of the Treaty of Rome concerned with employers and workers (users). These Regulations are administered in the UK by the HSC and the HSE. A list of relevant Directives is given at the end of this chapter.

The European Commission does not intend to create a European Community Health and Safety Executive to oversee and 'police' the implementation of health and safety Directives. To assist the authorities in the Member States to achieve full and equitable enforcement of Community safety legislation, the Commission works with a committee comprising officials from all Member States.

In the European Commission document *General Framework for Action by the Commission of the European Communities in the Field of Safety, Hygiene and Health Protection at Work (1994–2000)* (Nov 1993), the Commission recognized that a large amount of Community legislation had been introduced during the past few years in the field of safety and health at work. Therefore particular attention should now be paid to ensure consistent implementation of the adopted Directives by all Member States, with adequate control and monitoring procedures.

International law

International law is far from being a precise legal notion and usually lacks any fixed recognizable sanctions, but some agreements have begun to result from the efforts of supranational bodies, e.g. new rules for the health-related monitoring of airborne dusts and aerosols in the workplace have been recognized by the HSE. The Conventions which represent an important step towards the harmonization of dust-sampling practice worldwide are set out in the revised guidance document HSE 14, *Methods for the Deter-mination of Hazardous Substances* (frequently updated). The provisions represent target specifications for instruments used to assess the health effects of inhaled dusts and aerosols. Three aerosol fractions have been defined – inhalable, thoracic and respirable – and have been accepted by the International Standards Organisation, CEN and the American Conference of Government Industrial Hygienists.

Various forms of legislation

Bylaws, a form of delegated legislation, though limited in extent may be mandatory and can carry real sanctions. They are laws made by such bodies as Councils (e.g. Local Authorities) or Boards (e.g. the railway) under powers conferred by Acts of Parliament. Being merely local in application they are outside the scope of this chapter, but they must not be forgotten because specific Bylaws may dictate, for example, the legal method for the disposal of waste or an approved system of fire prevention. Details of Bylaws may be ascertained from Local Authorities.

The type of law considered so far has been Statute Law issuing from legislative authorities. Apart from this type of law there is Case Law based on precedents, that is, previous decisions of the Courts. Although the importance of Case Law has waned as UK Statute Law has gradually replaced the Common Law, the ancient unwritten law of the land embodied in judicial decisions, Case Law still has an important role in the interpretation of Acts and Regulations.

There are two divisions of law in practice – Criminal and Civil. Criminal Law implies an obligation which is enforced by the state through prosecution and the imposition of penalties. In the Criminal Court the Crown prosecutes the accused. The HSWA is first and foremost a criminal statute. The Civil Law deals with rights and duties between persons. These rights may arise from Statute Law, such as the Employment Protection (Consolidation) Act 1978, or they may be established by precedents in Case Law in respect of torts, e.g. negligence. In the Civil Court the Crown merely acts as an arbitrator between two contestants – plaintiff and defendant. The outcome of a civil case is not a penalty but a remedy, such as compensation for an injury. It is worth noting in passing that the HSWA expressly provides (s.47) that no action will lie for breach of the 'general duties' sections of the Act (s. 2–8), therefore compensation for injury caused by neglect of statutory duty under the HSWA, *per se*, is not possible.

The Health and Safety at Work etc. Act 1974

In the UK the HSWA is the principal legislation governing health, safety and welfare in the workplace. It was designed to promote greater cooperation and mutual responsibility amongst all those concerned with health and safety. Its obligations apply primarily to employers but it also applies to those in control of premises, contractors, the self-employed and employees. It is also stipulated that employers conduct their undertaking in 'such a way as to ensure so far as is reasonably practicable, that persons not in his employment who may be affected thereby are not thereby exposed to risks to their health and safety'. This implies that protection is extended not only to anyone entering an animal unit, e.g. a visiting scientist or a Home Office Inspector, but to any persons who avail themselves of the services of the animal facilities, such as an animal user receiving an infected animal.

The HSWA applies to all employers regardless of the numbers they employ and, indeed, to the self-employed (s.3). However, domestic servants are exempt. Furthermore, the Employers Health and Safety Policy Statement (Exception) Regulations 1975 exempted employers with less than five employees from having to prepare a health and safety policy.

The Act has established a comprehensive and coordinated system of control, inspection and enforcement. The broad approach of the HSWA is in contrast to the narrow 'checklist' approach of earlier legislation, e.g. the Factories Act 1961. There are four parts to the Act:

> *Part I* deals with health, safety and welfare at work, the control of dangerous substances and of certain emissions into the atmosphere. It is also concerned with the establishment of the HSC and the HSE.
> *Part II* deals with the Medical Advisory Service.
> *Part III* deals with building and construction.
> *Part IV* deals with miscellaneous and general matters.

The Act calls for the devising of safe systems of work and the identification and control of hazards as far as is reasonably practicable. These obligations of the HSWA are clarified in the Management of Health and Safety at Work Regulations 1992, as amended, in which employers are called upon to:

- assess the risk to the health and safety of their employees and to anyone else who may be affected by their activity, so that the necessary preventive measures can be identified;
- make arrangements for putting into practice measures following from risk assessment, covering planning, organization, control, monitoring and review;
- provide appropriate health surveillance of employees;
- appoint competent people to help devise and apply the necessary measures to comply with employer's duties;
- set up emergency procedures;
- give employees information about health and safety matters;
- cooperate with others on the same site;
- provide information to workers in their undertaking who are not their employees;

- make sure that employees have training in health and safety and are capable of avoiding risks;
- give sufficient information to temporary workers.

The employer cannot delegate his duty under the Act. He can delegate a health and safety management function, but the duty under section 2 of the Act remains with the employer.

Safety structures

The Health and Safety Commission (HSC)

The HSC is formed by representatives of employers and employees as well as those of local authorities. It has taken over from government departments the responsibility for producing and developing policies concerned with the health, safety and welfare of workers.

The Health and Safety Executive (HSE)

The HSE is a separate statutory body appointed by the HSC to work in accordance with directions and guidance given by the HSC. The HSE is responsible for the enforcement of legal requirements and the provision of an advisory service throughout industry, educational services and commerce.

Safety management, representatives and committees

The Management of Health and Safety at Work Regulations 1992 provide the framework for the application of the HSWA to safety management. The purpose of the Management of Health and Safety at Work Regulations 1992 (and guidance notes HS(G)65 etc.) is to clarify the requirements of the HSWA and in so doing to make it clear that managers, not safety officers, should manage health and safety. The management of health and safety should be a normal part of the line management function.

Reference to a safety policy occurs in section 2(3) of the HSWA. An employer of five or more employees must provide, in keeping with regulations under the Act, a written statement of health and safety policy and the organization and arrangements for its implementation. If requested by at least two safety representatives appointed by one or more trades unions, an employer must set up a safety committee. A notice of the composition of the safety committee and the area covered by it must be posted in the workplace. A copy of the minutes of the committee meetings should be pinned to appropriate notice boards or broadcast in some other suitable fashion. Communication on safety matters should be comprehensible and in some cases may need to be in another language as well as English. The safety policy statements must be revised when necessary and define levels of responsibility, naming the appropriate officials. The recording of accidents, injuries, incidents and medical treatment should be described.

Safety instructions should deal with the entire spectrum from hygiene to specific hazards such as animal handling, infection, chemical hazards and radiation. These 'in-house' codes can be based on guidelines devised by associated professional bodies. The HSE will advise on the compilation of such local rules which do not, however, have the force of ACOPs. Ideally there should be a specific set of rules for particular areas such as the animal house.

There is an important role for the elected safety representative in involvement with investigation of hazards, dangerous occurrences and accidents. It is the duty of the employer to keep the safety representative fully informed on all relevant matters. Safety representatives are entitled to inspect and take copies of any document which the employer is requested to keep by any relevant statutory provision, except confidential medical information.

Besides representing his fellow employees the safety representative should:

- consult with the employer on health and safety matters;
- investigate complaints by any employees he represents;
- present employees' complaints;
- make representations on general matters affecting the health, safety and welfare of the employees;
- carry out inspections in accordance with specified procedures.

In addition to safety officials, all other members of staff have duties with respect to health, safety and welfare, and these should be clearly outlined in the policy statement on safety. Heads of departments, managers and supervisors must be made aware of their obligations and should arrange consultations with their safety representatives to discuss, monitor, review and propose safety measures. A supervisor such as a technician in charge should:

- promote interest in health and safety;
- ensure compliance with safety rules;
- ensure proper maintenance of equipment;

- make safety equipment available;
- insist on the correct use of safety equipment.

'So far as is reasonably practicable'

Duties imposed by the HSWA are not absolute *per se*. As long as the employer acts reasonably in the circumstances, he does not automatically become guilty of an illegal act within the terms of the HSWA. The burden, however, is on the employer to show 'That it was not practicable or reasonably practicable to do more than was in fact done, or that there was no better practicable means than that which was actually used...'. The best practicable means are the best measures possible in the light of current knowledge and according to the availability of resources. 'Reasonably practicable' is less strict than 'practicable'. 'Reasonably' implies a computation of the balance between the risk and the sacrifice involved in carrying out the measures necessary to avert the risk, e.g. money, time or inconvenience. If, when this cost/benefit exercise has been carried out, the risk proves insignificant in relation to the sacrifice, it is not 'reasonably practicable' to take the steps necessary to control the hazard. Unfortunately, in practice such computations may turn on subjective judgements.

Health

Under the Management of Health and Safety at Work Regulations 1992 the topic of health surveillance merits special attention and must be an on-going process. Records are now required to be kept for 40 years. If the business activities of the employer cease, the records must be offered to the HSE. Detailed records should be kept of any employees exposed to toxic substances, biological hazards or radiation. Employees must be allowed access to their records and in turn must make themselves available for medical inspection and provide relevant information. The workplace must be open to inspection by the appropriate Medical Officer. Specific criteria for health surveillance are set out in Regulation 11 of the Control of Substances Hazardous to Health Regulations 1999 (COSHH). Further guidance on this matter is given in *Surveillance of People Exposed to Health Risks at Work 1990* (HS(G) 61).

Automatic pre-employment health screening is not compulsory. Occupational health provisions should include nursing arrangements, first aid and documentation of remedial action for contingencies, e.g. availability of antidotes to venom.

Employees must take care of their own health and that of others with whom they are involved at work (HSWA s.7). They must not recklessly interfere with anything provided in the interest of health and safety.

The Employment Medical Advisory Service (EMAS) is available in the UK for advice on occupational health problems.

Specific areas of health concern

Inhalation hazards. Where, for example, dust or fumes may constitute a hazard to health, engineering controls such as exhaust ventilation or enclosure of a process, and administrative controls such as exclusion of non-essential access and prohibition of eating in areas where hazardous substances may be present, must be considered first. COSHH makes it clear that the use of respiratory protective equipment is a last resort in controlling exposure to hazardous substances. When respirators are used they must, of course, be of a type fully suitable for the circumstances. See Chapters 5 and 7 for details of exposure limits.

Noise. The EEC Directive, the Protection of Workers from Noise (86/188/EEC), was implemented in the UK by the Noise at Work Regulations 1989. The Regulations stipulate three action levels: the first at a daily noise exposure of 85 dB(A) (roughly when normal conversation becomes difficult) and the second at 90 dB(A); the third is a peak sound pressure of 200 pascals.

The employer has to ensure that when any employees are likely to be exposed to the first action level or above, a competent person makes a noise assessment which is adequate for the purposes:

- of identifying which employees are so exposed, and
- of providing them with such information as will facilitate compliance with their duties under:

  Regulation 7 Reduction of noise exposure

  Regulation 8 Ear protection

  Regulation 9 Ear protection zones

  Regulation 10 Provision of information to employees

The noise assessment should be reviewed when:

- there is reason to suspect it is no longer valid, or

- there has been significant change in the work to which the assessment relates.

If the second or peak action levels are reached, measures must be taken to reduce noise exposure and the area must be designated an 'ear protection' zone. The demarcation sign is specified by BS 5378. Ear protectors must be provided and used. This legislation may be relevant to some cage-washing areas. Employers should encourage employees to have their hearing checked regularly. Any defects in equipment provided must be reported. The employer must inform the employees about the risk of hearing damage, the possible actions to minimize the risk, the availability of ear protection and the employee's own obligations.

Sight. The need to protect employees involved in light-intensive tasks and the provision of appropriate eye protection, e.g. for workers using lasers, is now dealt with in the Personal Protective Equipment at Work Regulations 1992.

Safety

Important factors in safety at work are:

- Provision and maintenance of plant and systems of work that are safe.
- Arrangements for the absence of risks in the use, handling, storage and transport of articles and substances.
- Provision of information, instruction, training and supervision to ensure safety.
- Maintenance of premises in a safe condition and the provision and maintenance of risk-free means of access and egress.
- Provision of a safe environment.

A definition of a safe system of work appears in HSE leaflet IND(G)76(L) 1992: 'A formal procedure which results from systematic examination of a task in order to identify all the hazards. It defines safe methods to ensure that hazards are eliminated or risks minimised'. Components that should be considered in association with a safe system of work include:

- Coordination of departments and activities.
- Proper layout of plant and appliances.
- Correct methods of using particular machines.

- Agreed methods for carrying out specific processes.
- Instruction of trainees and inexperienced workers.
- Regular sequence in which work is carried out.
- Provision of warning notices and special instructions.
- Procedures for introducing changes and explanation of the need for change.
- A contingency plan for dealing with emergencies.
- A monitoring regimen.

A safety policy must be wide enough to cover injuries or loss of health which may develop gradually over many years of exposure to a hazard or may appear long after the exposure has ceased. Implementation of Good Laboratory Practice could provide a basis for the formation of a good safe system of work.

Categories of special concern

Attention must be given to the training and supervision of, and assignment of tasks to, young persons aged under 18 years. Safety requirements with respect to disabled workers will take account of their disability. Where there are lone workers a reliable method of communication must be established, especially if they are in areas of high risk or in isolated units. An occupier owes a duty of care to the employees of others who are working on his premises.

Duties of employees

The HSWA (s.7) states the legal duties of employees in regard to health and safety at work. 'It shall be the duty of every employee while at work:

- To take reasonable care for the health and safety of himself and of other persons who may be affected by his acts or omission at work
- As regards any duty or requirement imposed on his employer or any other person by or under any of the relevant statutory provisions, to co-operate with him so far as is necessary to enable that duty or requirement to be performed or complied with.'

Workers ought consequently to point out dangers or shortcomings in the safety policy to management or the safety officer and, where prohibited, should refrain from such activities in the workplace as: eating, drinking, smoking, introducing naked flames. However, in the special circumstances associated with working in designated controlled areas, workers are permitted to drink from suitable water fountains (cf. Ionising Radiations Regulations 1985).

Training

An obligation to provide appropriate training flows from the provisions of the HSWA. To fulfil this duty the employer should:

- introduce comprehensive safety rules and procedures and induction training programmes for new recruits;
- provide repeat training at regular intervals;
- ensure that no employee transferred or promoted is permitted to start work in his new job until he has received training and instruction sufficient to enable him to perform the job without risk to health and safety;
- provide adequate training for managers;
- pay attention to the needs of existing employees;
- document details of training given and received.

The employee has a duty to accept instruction in safety hazards, the use of safety equipment and the correct use of all plant, materials and machines with which he is concerned. The training of Safety Representatives is a matter for their respective Trades Unions (cf. Safety Representatives and Safety Committees Regulations 1977).

The Health and Safety (Training for Employment) Regulations 1990 are concerned with incorporating those provided with relevant training (i.e. work experience whilst at school, etc.) within the meaning of 'employee' as defined in the HSWA.

Training provisions should be set out in the safety policy document. Specific areas of concern may demand special attention, for example, COSHH Regulations impose strict obligations in the matter of training in the handling of, and exposure to, hazardous substances and the precautions to be taken. Special training must be provided for instructors in this field.

Stress is laid by the Home Office on the importance of adequate and appropriate training in animal units. In the Code of Practice (Home Office 1989, p5 para 2.9) associated with the Animals (Scientific Procedures) Act 1986 there is the warning: 'Staff should be aware of the action to be taken in case of accident, fire or other emergencies, and of the potential existence of zoonotic organisms'.

The person named as holder of the Certificate of Designation must ensure adequate training of all personnel.

Information

The Health and Safety Information for Employees Regulations 1989 require employers to bring to the attention of employees information on requirements and duties under health and safety law, e.g. the local addresses of the relevant enforcing authorities, the HSE and the Employment Medical Advisory Service. If leaflets are used in place of posters, each employee must be supplied with a copy. Within the context of COSHH, employees should be kept informed of the results of the monitoring of dangerous substances. The HSWA and the Consumer Protection Act 1987 state that adequate information about the hazards or conditions for safe use of any article or substance being supplied, must be given to the user. This stipulation could have implications for animal suppliers.

Welfare

In the context of the HSWA, 'welfare' is concerned with the provision of toilets, washing facilities, clean working environment, etc. Welfare requirements are clearly set out, principally, in the Workplace (Health, Safety and Welfare) Regulations 1992, although reference is made to 'welfare' in a number of other Regulations.

Heat

Previous legislation in the form of the Factories Act 1961 (s.3) and the Offices, Shops and Railway Premises Act 1963 (s.6) had indicated 16°C as a minimum temperature for a workplace. These sections were repealed on 1/1/96 by the Workplace (Health, Safety and Welfare) Regulations 1992. Now the ACOP, L24 1992, which accompanies these Regulations sets out requirements for the maintenance of an appropriate temperature in all workplaces.

Light

Sufficient light must be available in the workplace so that eyesight is not jeopardized. In units without natural light, emergency lighting must be provided in case of power failure. An override switch should be provided in rooms where the lighting is time-controlled. If corridor lighting is time-controlled, there must always be sufficient lighting to allow safe movement when the main lights are off.

Lifting

The Manual Handling Operations Regulations 1992 implement in part the Directive (90/269/EEC) which is concerned with the manual handling of loads where there is a risk of back injury. It will no longer be sufficient to rely entirely on the weight of the load; factors such as individual ability to carry out the activity, the nature, form and size of the load, the type of activity and any necessary control measures should be considered.

There is a duty on the employer to:

- avoid, so far as is reasonably practicable, the need for his employee to undertake any manual handling operation at work which would involve a risk of their being injured. There is not a duty to avoid manual handling as such, so far as is reasonably practicable;
- assess all manual handling activities;
- reduce, at least as far as is reasonably practicable, any risks identified in the assessment.

Cleanliness

Employers have specific duties to provide, maintain and keep clean, washing and toilet facilities and to provide storage for clothing. Cleaning schedules should be introduced identifying:

- the item of plant, area and/or structure to be cleaned;
- the cleaning method, materials and equipment to be used;
- the frequency of the cleaning;
- individual responsibility for ensuring that the cleaning task is carried out;
- any precautions that may be necessary in the use of cleaning agents or electrical equipment.

The Code of Practice (Home Office 1989, p7 para 2.34) of the Animals (Scientific Procedures) Act 1986 refers to the need or the concern for welfare of animal staff. 'Personnel facilities should include staff and record rooms, sufficient changing rooms, decontamination areas, first aid and toilet facilities and space for storing protective and outdoor clothing etc.'

Hours of work

Excessive hours of work are a health and safety issue rather than a matter of welfare. It has been introduced at this juncture, as has the topic of termination of employment, as pertaining to general aspects of the HSWA.

Long hours and unsuitable shifts can cause physical and mental ill-health, induce fatigue, be a fertile source of accidents, to say nothing of oft-resulting domestic strife. In disputed cases about work-induced stress, the burden of proving that hours are not excessive nor shifts unsuitable could fall upon the employer. Rules concerning the hours of work must be stated in the written particulars available to an employee under the Employment Protection (Consolidation) Act 1978 (as amended). The recent European Working Time Directive (93/104/EEC) sets limits on weekly working hours and minimum rest periods for most areas of employment and is dealt with in the UK in the Working Time Regulations 1998. The Health and Safety (Young Persons) Regulations 1997 are also relevant.

Termination of employment

Unsafe conduct by an employee may warrant termination of the employee's contract of employment. To be considered as 'fair', the termination should conform to the statutory rules relating to unfair dismissal. The employer's action must be reasonable in the circumstances.

Enforcement

The HSWA may be enforced by the HSE or Local Authorities (cf. the Health and Safety (Enforcing Authority) Regulations 1998). A duly appointed inspector may enter premises at any reasonable time with any necessary equipment, e.g. for monitoring conditions. They may examine anything on the premises and take such action as they deem necessary. They are empowered to require relevant information from any person.

An improvement notice may be issued if there is a contravention of the law. Whether or not one is issued is a matter for an inspector's discretion. Section 22 of the HSWA which outlines an inspector's powers to serve a prohibition notice was amended by the Consumer Protection Act 1987. An inspector may serve a notice where he is of the opinion that there is a serious risk of personal injury. Prosecution following failure to comply with a notice will be of the person (whether an individual or a Limited Company, etc.) on whom the notice was served.

Specific hazards in the animal facility

'Hazard' – which is a potential danger, avoidable or unavoidable in certain circumstances – should not be confused with 'risk'. A 'risk' means that there is a reasonable probability of an unfavourable outcome. Risk assessment should contain an element of quantification, a probability measurement.

Animals

The Code of Practice (Home Office 1989, p6 para 2.20) of the Animals (Scientific Procedures) Act 1986 draws attention to specific hazards associated with animals. 'Precautions should be taken in animal rooms to minimise the exposure of personnel to hazards which may arise from the incorrect handling of animals, such as bites and scratches, allergens and infections and to prevent exposure to hazardous treatments intended for, or applied to, animals.'

Physical hazards

The likelihood of damage from butting, stamping or biting, etc., should be considered as part of the overall risk assessment of the animal facility. Consequent on such an assessment there should be an agreed system of work and availability, if need be, of protective clothing as well as equipment and drugs for restraining purposes. Handling methods should balance minimum effective restraint and minimum risks. All those involved with animals must be trained in methods of handling and correct use of methods of restraint. Personal licensee training in keeping with the Directive (86/609/EEC) and associated with the Animals (Scientific Procedures) Act 1986 provides the opportunity for this type of instruction.

In-house rules should be drawn up for dealing with large and dangerous animals, in particular monkeys and venomous snakes. Any injury by an animal, however slight, must be reported immediately and appropriate action taken. The Dangerous Wild Animals Act 1976 refers to non-domesticated species considered capable of causing greater injury than a domestic animal. Keepers of such animals must hold a local authority licence. Zoos and premises which are registered with the Home Office are exempted from the provisions of this Act.

Zoonoses

The Agriculture (Miscellaneous Provisions) Act 1968 introduced control of zoonoses. The Zoonoses Order 1975, as amended, exempts those involved in research from the requirement to report occurrences of brucellosis and salmonellosis in food animals and has updated the list of dangerous organisms.

The COSHH assessment (Control of Substances Hazardous to Health Regulations 1999) should take account of all possible zoonotic hazards resulting from the presence of microorganisms hazardous to health. Details of such pathogens can be found in Chapter 3 of this book. The consequences of the assessment should be strict control of contact with possibly infected animals, prophylactic procedures where possible, e.g. vaccination, and strict adherence to quarantine rules. Tetanus immunization must be kept up-to-date and regular medical checks must be available to endangered staff.

The Code of Practice (Home Office 1989, p21 para 3.77) of the Animals (Scientific Procedures) Act 1986 refers to this hazard: 'Animals that may harbour zoonotic agents should be caged, managed and handled in such a way as to minimise any risk of infection being transmitted.' Advisory Committee on Dangerous Pathogens (ACDP) 1998 covers infection risks from simians.

Quarantine

Under the Rabies (Importation of Dogs, Cats and other Mammals) Order 1974 there is a compulsory six-month quarantine for most animals entering the UK. Perpetual quarantine is imposed in the case of vampire bats. In the case of birds the stipulated period of quarantine is 35 days. In specific cases the quarantine period will be indicated in the import licence. Details about licences, health certificates, rabies and other quarantine requirements should be obtained from the Animal Health Division, MAFF, or the Department of Agriculture for Scotland or from the Wildlife and Conservation Licensing Section, Department of the Environment, Bristol. In Northern Ireland importation is controlled by the Department of Agriculture.

Quarantine must be spent at premises authorized by a MAFF Inspector. Strict and detailed rules are set down for these facilities, which are subject to regular veterinary inspection. Permission must be obtained for the movement of a quarantined animal and any vehicle used must be approved by MAFF.

European Directives relevant to the importation of possibly infected animals and animal products have been issued. Three are of special significance:

- Directive 90/425/EEC relates to animal pathogens.
- Directive 92/60/EEC is intended to eliminate veterinary and zootechnical checks at internal borders of the EU. This is to be achieved by imposing the obligation of performing the checks at the place of despatch of the animals and/or animal products.
- Directive 92/65/EEC, known as the Balai (Catch-all) Directive, controls the importation from outside the EU into any Member State of animals, semen, ova and embryos. In doing so it makes it unnecessary for full rabies quarantine to be imposed on laboratory animals originating from EU sources, providing such sources are registered with the relevant ministry in the source country.

These directives have been implemented by the UK in the Animal and Animal Products (Import and Export) Regulations 1993 and 1995. The Regulations are executed and enforced by the local authority (Reg 3). Schedule 2 names the border inspection points for the reception of animals from countries outside the EU which are named in various European Directives. Notwithstanding this new legislation, the Rabies (Importation of Dogs, Cats and Other Mammals) Order 1974 continues to apply to all carnivores, primates and bats. It shall continue to apply to the importation of all other animals unless such animals are imported by way of trade, e.g. dogs and cats for breeding, and can be shown to have been born on the holding of origin and kept in captivity since birth (Sch 6).

Industrial diseases

Some animal-related diseases are referred to in the Industrial Diseases (Notification) Act 1981. Under the Social Security (Industrial Injuries) (Prescribed Diseases) Regulations 1985 there are prescribed occupational conditions or diseases which will qualify claimants to receive benefits. Some of these may be of relevance to workers with farm animals. The list includes: anthrax infection, avian chlamydiosis, hydatidosis, leptospirosis, *Brucella* infections, *Streptococcus suis* infection, ovine chlamydiosis and tuberculosis.

Allergies

Laboratory animal allergy (LAA) is a recognized clinical syndrome and can render a person unfit for work with animals (see Chapter 2). The Reporting of Injuries, Diseases and Dangerous Occurrences Regulations 1995 lists asthma as a reportable disease if it is a result of work exposure. Some of the causal agents listed in the Regulations may be associated with work in an animal unit, e.g. animals, enzymes, antibiotics and glutaraldehyde. Health surveillance of staff should identify symptoms of allergy at an early stage. An effective programme of control of allergens should be instituted, such as appropriate ventilation and the provision of personal protective equipment, though the latter must be regarded as a second line of defence.

Other hazards

Pathogens

The Biological Agents Directive (90/679/EEC) defines 'biological agents' as microorganisms including genetically modified organisms and cell cultures, and applies to work activities where employees may be exposed to such agents. The COSHH Regulations 1999 implement the European Directives on the subject. The COSHH Regulations control the use of biological agents in the workplace and define more clearly the following terms:

> *Biological agent* – any microorganism, cell culture or human endoparasite, including any which has been genetically modified, which may cause infection, allergy, toxicity or otherwise create a hazard to human health.
>
> *Microorganism* – a microbiological entity, cellular or non-cellular, which is capable of replication or of transferring genetic material.

The classification and additions of newly recognized biological agents is an ongoing process. Amendment (93/88/EEC) of the Classification of Biological Agents Directive is reflected in amended COSHH Regulations.

Microbiological hazards are dealt with in detail elsewhere in this book (Chapter 3). What follows here is merely an outline of the type of action demanded from employers. There must be an assessment of the risks of any work with biological agents likely to be hazardous to the health of employees. A distinction is made between workers who work directly with biological agents and workers whose exposure may be incidental, e.g. maintenance staff. Hazards may be removed by substituting a non-pathogenic organism for a harmful one or treating organisms in such a way as to render them harmless. Control of exposure should be secured, where reasonably practicable, by means other than by personal protective equipment. The use of safety cabinets and isolators will sometimes be necessary. Specific guidance is given in the Advisory Committee on Dangerous Pathogens (ACDP) *Categorisation of Biological Agents according to Hazard and Categories of Containment* (Health and Safety Commission 1995) which describes four levels of laboratory containment corresponding to the four hazard groups. A more recent publication is ACDP (1997) on the same subject.

Employers are required to take all reasonable steps to ensure that control measures are complied with and that equipment is examined and maintained in an efficient state. Records of examinations and tests must be kept for at least five years. Employees are required to make use of control measures and report any defects.

Exposure of employees to biological hazards must be monitored and records for identified employees must be kept for 40 years. There are no occupational exposure limits for work with micro-organisms. Health surveillance must be provided where:

- an identifiable disease may be related to exposure;
- there is a reasonable likelihood that the disease will occur under conditions of work;
- there are valid techniques for detecting disease.

The Reporting of Injuries, Diseases and Dangerous Occurrences Regulations 1995 and amendments require reports to be made to the HSE of the death of any person as a result of exposure to biological hazards arising out of or connected with work and any potentially serious accident or incident, e.g. needle-stick injury, where a pathogen is involved. There should be emergency plans for dealing with accidents with microbiological hazards. A new ACOP (L5) on the control of biological agents based on the 1999 COSHH Regulations was issued in 1999. The Rabies Virus Order 1979 permits the importation, use and keeping of rabies virus only under licence.

Storage, transport and disposal of biologically hazardous material. Pathogens must be stored safely and securely in robust containers with adequate labelling. Appropriate Regulations must be complied with. Incineration, after autoclaving, is the preferred mode of disposal. Guidance from the HSC (1991) appears in *Safe working and the prevention of infection in clinical laboratories*.

Postage of pathological material. Postage of pathological material must comply with the Post Office's leaflets k681 (inland mail) or DSO 61 (overseas). The former sets out the requirements for packing such material, including the use of leak-proof containers and absorbent material. In some cases the use of polystyrene boxes may be appropriate. The package must be marked 'fragile with care' and 'pathological specimen' and sent by first class mail.

Carcinogens

The ACOP (1999) *Control of Carcinogenic Substances* is intended to facilitate in practice the observance of the COSHH Regulations 1999 and implements in the UK the provisions of Directive 90/394/EEC and amendments, for the protection of workers from exposure to carcinogens (substances that may cause cancer). There must now be regular assessments at least every five years of risks associated with exposure to carcinogens. Formerly such an assessment was demanded only when the nature of the work changed. A new risk phrase 'may cause cancer by inhalation' has been introduced for use in assessment reports. Monitoring and health surveillance records relating to carcinogenic substances are required to be kept for 40 years. This extended period of 40 years also now applies to the keeping of monitoring and health surveillance records for all hazardous substances.

Legal control of the presence of carcinogens is of constant concern. This state of affairs is reflected in two amendments to European Directive 90/394 EEC (97/42/EC and 98/C 123/12). The original Directive (90/394/EEC) established minimum standards across the EU for the prevention and control of occupational exposure to carcinogens and provided for the setting of binding limit values (exposure limits) for carcinogens at EU level. The 1st Amendment revised the European exposure level for benzene. It dealt with the inconsistencies and ambiguities in the application of the original Directive. It added types of chemical previously excluded such as formulated pesticides, cosmetics, medicines and hazardous waste. One result of the 1st Amendment has been that in future the list of altered Maximum Exposure Limits (MELs) will be removed from Schedule 1 of COSHH. The HSC will approve future changes to the list and they will be published in the HSE's annual publication EH 40 — *Occupational exposure limits*. The phased implementation of Regulations arising from this 1st Amendment is due by 26 June 2003.

The 2nd Amendment to the Carcinogen Directive is currently under negotiation (in 1999). It is intended to extend the provisions of the Directive to substances which cause heritable genetic damage – i.e. meet the criteria for classification as Category 1 or 2 mutagens as defined in the Dangerous Substances Directive (67/548/EEC).

Genetic modification

The Directives on the Contained Use of Genetically Modified Microorganisms (90/219/EEC under A.130S) and on the Deliberate Release into the Environment of Genetically Modified Organisms (90/220/EEC under A.100A) are concerned with human health and safety, the protection of the environment and affect facilities working with transgenic animals (see Chapter 4). The two Directives have been implemented in the UK by Regulations under the HSWA and the Environmental Protection Act 1990 as the Genetically Modified Organisms (Contained Use) Regulations 1992 (as amended) and the Genetically Modified Organisms (Deliberate Release) Regulations 1992 and 1995. See Chapter 4 for more information.

Notification of the first use of premises for genetically modified organisms is required but even then consent from the HSE must be obtained for certain activities. Provisions must be made for emergencies and accidents must be duly notified. Accidents and serious incidents will be the concern of the Genetic Modification Safety Committee. Environment risk assessments are required, where relevant, as well as assessments of risks to human health and safety. There is a demand for the use of the best available technology to prevent environmental damage. Inspectors have the power to issue prohibition notices and courts are able to order remedies. The Secretary of State has the power to remedy harm at the expense of convicted persons. There is an Advisory Committee on release to the Environment. The Genetically Modified Organisms (Contained Use) (Amendment) Regulations 1998 came into force recently, but the situation is still fluid. The Council Directive of 26 October 1998 (98/81/EC) amends Directive 90/219/EEC. Regulations will be made implementing this new Directive by 30 April 2000.

Hazardous substances

Hazardous substances are dealt with comprehensively in COSHH Regulations. Substances hazardous to health encountered in the animal unit could include cleaning fluids, disinfectants, anaesthetics, allergens and pathogens.

A main feature of COSHH is the assessment of health risks. Before starting any work which is liable to expose employees to substances hazardous to health, there must be a suitable and sufficient assessment of the risks which may be created by that work and of the steps needed to comply with the legal requirements of COSHH. A review of the assessment will be required if its validity becomes suspect or if changes occur in the workplace. Assessments need to be made by a competent person. 'Competent person' is defined in the Carcinogen Directive (90/394/EEC) as 'any person who has the necessary knowledge, experience, practical ability and skills to perform the task in question'. The potential effects of substances must be considered with respect to their quantity and form, and the amount to which employees may be exposed. Such exposure should be compared to any published standards. Assessors should take into account past experiences, previous records and available information on toxicity.

COSHH regulations cover virtually all substances causing adverse health effects at work. There are five identified categories:

- Substances in the Authorised and Approved list within the Chemicals (Hazard Information and Packaging for Supply) Regulations 1994, in the very toxic, toxic, harmful, corrosive or irritant categories.
- Substances with a Maximum Exposure Limit specified previously in Schedule 1 of COSHH and henceforth in EH 40, or substances for which the HSC has approved an Occupational Exposure Standard.
- Microorganisms creating health hazards.
- Any dust at a substantial concentration in air.
- Any substance which creates a health hazard like any of the above-mentioned.

The most recent revision of COSHH has been the Control of Substances Hazardous to Health Regulations 1999. Since COSHH 1994, amendments to these Regulations have implemented the Commission Directive 96/55/EC and many new substances were brought within the scope of COSHH. A new aspiration hazard risk phrase (R65) was introduced. There are revised criteria for respiratory irritants and sensitizers. Test methods are described for the determination of the hazardous properties of chemicals. Legislation in this area does not stand still. A Council Directive (98/24/EC) on the protection of the health and safety of workers from the risks related to chemical agents at work was adopted on 7 April 1998 and should produce Regulations binding citizens of the UK by 5 May 2001.

Control of hazardous substances by means of the provision of personal protective equipment is not permissible unless the employer can show that exposure cannot be prevented or controlled by other means. There are two sets of Occupational Exposure Limits (OELs) expressed as concentrations of hazardous substances in the air averaged over a specified period, referred to as a time-weighted average (TWA). Two time periods are used, long-term (8 hours) and short-term (15 minutes). Short-term exposure limits (STELs) are intended to prevent damage such as eye irritation, which may be caused by a few minutes' exposure. Maximum Exposure Limits (MELs) are set for substances liable to cause serious health effects such as cancer or asthma, and for which safe levels of exposure cannot be fully determined or for substances for which there are safe levels but control to those levels is not reasonably practicable. Occupational Exposure Standards (OESs) are set at levels that will not damage the health of workers exposed daily by inhalation. OELs and MELs are reviewed continually and published in EH 40.

The duty of employers as regards substances assigned a MEL is to reduce the level of exposure to at least below the MEL and as low as reasonably practicable. Special attention must be paid to substances for which there are no apparent thresholds below which adverse effects do not occur or where relatively high concentrations of a substance over very short periods may be injurious. Although in exceptional circumstances OESs may be exceeded, the employer must identify the reasons for the values being exceeded and make every reasonably practicable effort to reduce exposure down to the acceptable limits at the earliest opportunity (see Chapters 5 and 7).

Substances that can be absorbed through unbroken skin are given an 'Sk' notation. In risk assessment, consideration should be given to the presence of other factors affecting how people respond to exposure to hazardous substances. Such factors could be alcohol, drugs, high temperatures and varying work patterns. Where there is exposure to several substances, the overall effect of the mixed exposures may be more hazardous than the sum of the exposures separately. In risk assessment the presumption should be that the effects are additive. In such cases it is recommended that a specialist be consulted.

CHANs (Chemical Hazard Alert Notices) also appear annually in EH 40. They are issued for substances where current scientific information indicates it is not possible confidently to set a safe exposure limit. The setting of MELs for these substances will be considered by the HSC's Advisory Committee on Toxic Substances (ACTS).

Notification of new substances

The Notification of New Substances Regulations (NONS) 1993 implemented several Directives, e.g. Risk Assessment Directive (93/67/EEC) aimed at providing greater protection for people and the environment from the possible ill-effects of new substances. Manufacturers and importers are required to provide information (called a 'notification') on the properties, uses and quantities of any new substances they intend to place on the market.

Transport, packaging and storage

The Chemical (Hazard Information and Packaging for Supply) Regulations 1994, known as CHIP 2, implemented nine Directives, e.g. Dangerous Substance Directive (67/548/EEC). The CHIP 2 Regulations require suppliers to provide a safety data sheet and to label packages in the format set out in the Regulations, which require comprehensive hazard information. In association with CHIP 2 there is a guidance ACOP (L63 1993), an approved supply list (L115 1998), and an ACOP of safety data sheets (L62 1995). These publications are revised from time to time.

Noxious gases

ACOP 30, *Control of Substances Hazardous to Health in Fumigation Operations* (L86 1996), specifies the measures which need to be taken to protect employees. Only those specially trained and conversant with the facility's ventilation system should undertake fumigation. Provision of personal protective equipment is allowed only where the employer is able to show that prevention of exposure to the fumigant is not reasonably practicable nor possible by other means.

Due care must be taken with other gases, such as carbon dioxide or liquid nitrogen, which may be used in the animal unit.

Medicines, drugs and poisons

The COSHH Regulations do not apply to patients given substances for medical reasons (including clinical trials), but staff involved in the administration (pharmacists, nurses, doctors, etc.) are covered by the Regulations. Veterinary surgeons and researchers who administer substances to animals for treatment or clinical trials are also covered by COSHH Regulations. According to the provisions of the Prescription Only Medicines (Human Use) Order 1997, prescription-only medicines may be supplied to universities or other institutions concerned with research. The Misuse of Drugs Act 1971 lists and classifies controlled drugs and lays down restrictions on the importation, production, supply and possession of such drugs. The Misuse of Drugs (Safe Custody) Regulations 1973 stipulates that where any controlled drug (other than a drug specified in Schedule 1 of the Regulations) is kept otherwise than in a locked safe, cabinet or room which is so constructed and maintained as to prevent unauthorized access to the drug, any person having possession of the drug shall ensure that it is kept in a locked receptacle which can only be opened by an authorized person. Part III of these Regulations specifies the details of documentation and records.

The administration of dangerous drugs to animals, e.g. volatile anaesthetics, may be hazardous to the operator, so appropriate protective measures must be in place (see Chapter 5). Health Services Sheet No. 7, *Waste Anaesthetic Gases*, provides practical guidance on control of exposure and on how to achieve compliance with legal duties. The gases covered by Sheet No. 7 are nitrous oxide, halothane, enflurane and isoflurane, which now have OESs (see EH 40).

Poisonous substances that are not medicines are controlled by the Poisons Act 1972 and the Poisons Rules 1982, which restrict the supply of substances listed in the Poisons Lists Order 1978. Poisons must be sold through pharmacies, except for Part II poisons which may be supplied by listed sellers to a person or institute concerned with education or research. There are special requirements for the labelling, secure storage, responsible handling and transport of poisons.

Radiation hazards

Details of these hazards are dealt with in Chapter 6 of this book. The principal forms of legislation are Radioactive Substances Acts 1948, 1960 and 1993. The last Act consolidates and amends the previous legislation. Numerous Regulations and codes have been issued in connection with these Acts, of which the Ionising Radiations Regulations 1985 are a relevant statutory provision of the HSWA made under s.15. They set out the duties relating to the safe use of ionizing radiation. The primary duty under Regulation 6 is: 'Every employer shall, in relation to any work with ionising radiation that he undertakes, take all necessary steps to restrict so far as reasonably practicable the extent to which his employees and other persons are exposed to ionising radiation'.

The means recommended to avoid hazards include the provision of appropriate facilities and extensive training of all involved with radioactive substances. Intensive monitoring and efficient health surveillance are essential. Strict recording is require in regard to the use, leakage, storage and disposal of radioactive substances.

The Radioactive Substances (Substances of Low Activity) Exemption (Amendment) Order 1992 excludes from statutory controls the disposal of radioactive waste containing carbon 14 or tritium, the activity of which does not exceed 4 becquerels per ml.

The Ionising Radiation (Outside Workers) Regulations 1993 require an exchange of information between employers of peripatetic workers and the operators of the sites where they work, to ensure that only suitable workers are assigned to work with radiation. The relevant passbook must contain the information detailed in the Schedule to the Regulations.

Legislation concerning the hazards of ionizing radiation in the workplace has moved on apace in Europe. Council Directive 96/29/Euratom revised previous directives and was adopted on 13 May 1996 for implementation on 13 May 2000. The Directive strengthens controls on ionizing radiation and reduces the maximum levels of exposures to radiation doses. It introduces requirements regarding exposure to natural radiation and emergency preparedness. Regulations arising from this directive may be in place in the UK in 1999/2000 and will probably incorporate the Ionising Radiation (Outside Workers) Regulations 1993 and implement provisions of the medical exposures directive. Provisions for emergency preparedness in the directive 96/29/Euratom will be implemented in separate regulations which will subsume the Public Information for Radiation Emergencies Regulations 1992.

Visual display units

The Health and Safety (Display Screen Equipment) Regulations 1992 give partial effect to Directive (90/270/EEC) on the minimum safety and health requirements for work with display screen equipment. The minimum requirements laid down in the Directive are produced in the Schedule to the Regulations. The Regulations require assessments of risk such as muscular problems, eye fatigue and mental stress. Steps should be taken to reduce hazards by ergonomic design of equipment. Free eye tests should be made available. Individuals with a history of photosensitive epilepsy or who suffer skin rashes apparently connected with a VDU should seek advice from the Employment Medical Advisory Service (EMAS).

The Management of Health and Safety at Work (Amendment) Regulations 1994, which implements Annexes I and II of Directive 92/85/EEC, is relevant in this context as it deals specifically with the health and welfare of pregnant women in the workplace. Important publications on VDU use are *Health and Safety (Display Screen Equipment) Regulations 1992, Guidance on Regulations* (L26 1992) and *VDUs. An easy guide to the Regulations* (HS(G) 90 1997). Both documents are available from the HSE.

Regulations controlling the use of VDUs may be of marginal concern in an animal facility since the definition of a VDU user, within the legislation, is: one who habitually uses display screen equipment as a significant part of their normal work.

Equipment

Machinery

Under the HSWA as amended by the Consumer Protection Act 1987, manufacturers, designers, importers and suppliers of articles for use at work have a duty 'to ensure, so far as is reasonably practicable, that the article is so designed and constructed that it will be safe and without risks to health at all times when it is being set, used, cleaned or maintained by a person at work'.

More direct obligations concerning work equipment were laid upon the employer by the Provision and Use of Work Equipment Regulations 1992. These Regulations give partial effect to Directive (89/655/EEC) which provides for legislation with respect to equipment at work. The Regulations give direction on the suitability, maintenance, cleaning, inspection and stability of equipment. Efficient control systems, methods of guarding, particularly dangerous parts of machines, and warning markings are demanded. Sufficient training and instruction, with stress on safety, in the use of machinery must be given. The working conditions in which machinery operates must be considered. The regulations were revised in 1998 and the HSE has published official guidance in an ACOP (L22 1998).

Directive 89/655/EEC, on the Minimum Safety and Health Requirements for the Use of Work Equipment at Work, was amended in 95/63/EC. Changes are concerned with the following aspects of equipment safety:

- minimum requirements for the provision of physical safeguards for mobile equipment and lifting equipment;
- a generally stated requirement for initial and in-service inspections, which would give Member States freedom to continue their existing regimens where these exist.

Equipment in the animal unit. Caging and bottles can be a source of danger: poor finishing and projections must be avoided, damaged cages should be discarded or repaired at once, plastic rather than glass bottles ought to be considered.

In every well-equipped animal facility there will be an abundance of machinery which is required to be assessed in keeping with the Provision and Use of Work Equipment Regulations 1992, e.g. cleaning and polishing machines, incinerators, autoclaves and washing machines. (It may be relevant to note here that the Public Health Act 1936 prohibits passing into public sewers waste steam or any liquid with a temperature higher than 43°C.) Some items of plant may be subject to the Pressure Systems and Transportable Gas Containers Regulations 1989, e.g. steam generators, autoclaves and some disinfecting machinery.

Attention should be given to the positioning of large machines. Factors to be considered in this matter are space, light, ventilation and accessibility. Risks associated with smaller machines for such activities as tattooing, shearing or bottle-washing should not be ignored.

In large establishments, the Lifting Plant and Equipment (Records of Test and Examination etc.) Regulations 1992 may demand attention. A guide to the Regulations is available from the HSE (L20 1992).

Electricity

Electrical safety in all work activities has been legislated for in the HSWA. The clarification of an employer's duties in this area is to be found in the Electricity at Work Regulations 1989. A Memorandum of Guidance to the Regulations has been published by the HSE (HS(R) 25) 1989.

Construction of equipment must be to EU-accepted good engineering practice standards to ensure that it is safe when connected to the electricity supply and provides an acceptable level of protection against shock. All flexible cables and cords must be to the required safety standard. All relevant warning signs must be clearly displayed on equipment.

All electrical systems shall be constructed and maintained to prevent danger. Work and maintenance must be carried out in accordance with established principles of safety. Electrical equipment must be suitable and used only within its limits and in the environment for which it is designed. Only properly trained and competent persons or those under appropriate supervision should be allowed to service electrical installations. There must always be means for interrupting the supply of electricity to equipment. Adequate work space must be provided around electrical equipment.

Testing. Inspection should be carried out in accordance with the regulations for electrical installations produced by the Institution of Electrical Engineers. There must be an efficient system of defect-reporting and reliable record-keeping. There should be visual inspection of cables, plugs and equipment before use. Class I (earthed) portable electrical equipment should receive a detailed inspection once a year. The frequency of a detailed inspection of Class II (double-insulated) portable electrical equipment depends on usage.

Electricity in the animal house. Special care must be taken regarding electricity in rooms housing aquaria and amphibia. Plugs in rooms that are regularly washed down must be splash-proof and earth leakage protected. Where electrophoresis tanks operate at voltages between 120V DC and 650V DC, they should be placed in an interlocked enclosure so that the current goes off if the lid is opened. If the voltage is above 650V DC an additional safety switch should be fitted to control a DC output earth dumping switch.

Personal protective equipment

The supply by the employer to the employee of necessary and appropriate personal protective equipment is mandatory under the Personal Protective Equipment at Work Regulations 1992. These regulations give effect to the Directive (89/656/EEC) on the minimum health and safety requirements with regard to the use by workers of personal protective equipment at work. The Regulations require employers to:

- provide personal protective equipment (PPE) where risks to health and safety cannot be controlled adequately by other means;
- select PPE that is suitable for the risks to be protected against;
- maintain PPE to acceptable standards and provide storage space;

- ensure that the PPE provided is properly used;
- ensure employees are given information on and instruction in the use of the PPE supplied.

Monitoring of exposure and provision of health surveillance should continue, where appropriate, alongside the use of PPE. Employees have a duty to make proper use of the PPE and report any losses or defects. No charge may be levied on an employee for the use of required PPE.

The Personal Protective Equipment (EC Directive) (Amendment) Regulations 1994 implemented Directives 93/68/EEC and 93/95/EEC. These Regulations insist that all PPE must be duly marked with 'CE marking' indicating conformity with European standards. The Personal Protective Equipment (EC Directive) (Amendment) Regulations 1996 implemented the European Directive (96/58/EC) on this topic and provided for the omission of certain 'Additional Information' from CE conformity marking on personal protective equipment.

The workplace

The demand for attention to health and safety in the workplace expressed in the HSWA is fully detailed in the Workplace (Health, Safety and Welfare) Regulations 1992. These Regulations implement Directive 89/654/EEC on minimum workplace standards. The provisions of the regulations fall into three categories:

- The initial structure of the workplace in relation to the user, expressed in design features geared to safety, e.g. providing protection against falling objects.
- Building/worker interactions; providing an appropriate layout, sufficient ventilation, adequate space, convenient work stations, suitable temperature (cf. guidance provided in the ACOP, L24 1992), appropriate lighting.
- Basic facilities, such as toilets, washing, etc., as set out in Regulations 10 and 11.

Many of these amenities were already called for under previous legislation. New provisions include safe window cleaning, the protection of non-smokers in 'rest areas' and rest facilities for pregnant or nursing mothers.

The duty to provide safe means of access and egress contained in the HSWA is stressed in the Regulations, e.g. requirements for safe traffic routes and doors and gates of safe construction. Particularly pertinent to the animal unit are provisions for the cleanliness of all surfaces and the stipulations that all passageways and stairways must be free of obstructions and warnings must be given of slippery floors.

The stores in the animal house need special attention. The consequence of unplanned storage can be catastrophic. Hazards can arise from badly stacked hay, poorly packed disinfectants and the mixing of chemicals harmless when on their own but hazardous in the presence of other specific chemicals. Employers have a duty under COSHH to take cognisance of hazards associated with store rooms. Storage arrangements should form part of the assessment required under Regulation 6.

Fire

The Fire Precautions (Workplace) Regulations 1997 implement the safety provisions regarding fire of the 'Framework' Directive (89/391/EEC) and the Workplace Directive (89/654/EEC). The Home Office guidance notes (1997) accompanying the Regulations stress the need for risk assessment based on the number of people present, structural features, the flammability of furnishings and the nature of the work.

Stocks of highly flammable liquids must be kept to a minimum and stored in a fire-resistant container. If held in a refrigerator the equipment must be spark-proof. Flammable liquids should be used only in a fume cupboard away from any machinery that might cause ignition. Carcasses of animals killed with volatile anaesthetics must be disposed of carefully according to best practice.

The fire certificate

Under the Fire Precautions Act 1971, administered by the Home Office and the Fire Authorities, work premises where 20 or more people are employed or 10 workers are situated other than on the ground floor, require a fire certificate. To obtain a fire certificate an occupier or owner must complete application form FPI (rev). The Fire Authority inspects the premises to ascertain the adequacy of:

- the means of escape;
- the methods of ensuring that the means of escape can always be used;
- the means of fighting fire;
- the fire-fighting system.

According to the Fire Precautions Act 1971, in the interim period between application and issuing the certificate, occupiers have the duty to:

- ensure the means of escape can be used;
- train employees on the procedures to be followed in case of fire;
- maintain existing fire-fighting equipment.

The fire certificate may impose specific conditions and must be displayed on the premises. Proposed alterations to buildings and of the siting of dangerous substances must be notified. Improvement or prohibition notices may be issued, or a fire certificate withdrawn for non-compliance.

Fire alarms

All new fire alarms should comply with BS 5839 Part I 1988 'Code of Practice for system design, installation and servicing'. Under most circumstances, equipment used must comply with BS 5445 'Components of automatic fire detection systems'. With both manual and automatic fire alarm systems, it is important that regular testing, inspection and maintenance are carried out.

The COP (Home Office 1989, p11 para 2.64) of the Animals (Scientific Procedures) Act 1986 recommends fire alarms which are inaudible to small rodents. In Condition 15 the Certificate of Designation stipulates that adequate precautions against fire shall be maintained at all times.

Means of escape

Escape routes should be checked daily to ensure that fire exit doors are not locked, that fire doors are not wedged open and that all signs are in place and visible. Emergency lighting systems, if necessary, must be installed and maintained in accordance with BS 5226 'Emergency lighting'.

Fire drills are recommended to take place at least once a year. Records should be kept indicating the date, evacuation time and the number of participants, etc. Whatever the number of staff, the responsibility for action in the event of fire should be assigned to named persons.

In case of fire, the fire fighters will normally attempt to isolate the fire and will aim to save human life first. No attempt should be made to move animals unless the person in charge who is familiar with the animals concerned is present. The fire fighters will not enter isolation units where work on dangerous pathogens is being carried out except to rescue personnel, neither will they open refrigerated containers holding dangerous pathogens. Such containers should be clearly marked. Fire Authorities should be made aware of any special features, e.g. an aquarium or the presence of venomous snakes, or any particular type of work which could influence the approach of fire fighters to a conflagration.

Control of other hazards and risks

Security

Security has become associated with safety in the animal house because of threats from those opposed to the use of animals in research. Protection of staff must be high on the agenda of animal house management. Possible hazards arising out of the present fraught situation are: personal assault, injury from car or letter bombs, fear and damage to personal property. The COP (Home Office 1989, p5 para 2.12) of the Animals (Scientific Procedures) Act 1986 recognizes the need for security: 'Advice should be taken about security from Crime Prevention Officers from the local police or other experts during the design of new facilities or modifications of existing premises'.

The law allows the use of reasonable force to evict intruders who refuse to leave when requested, or to prevent crime. Necessary force can be used in self defence. Unfortunately these terms lack precision in present law, and court decisions in such cases are not always predictable. However, the keeper of a dangerous animal shall not be liable for injuries to a trespassing plaintiff if the animal was not kept on the premises for the protection of persons or property or, if it were kept there for those purposes, such keeping was not unreasonable (cf. the Animals Act 1971). The Guard Dogs Act 1975 requires that guard dogs, while protecting commercial premises, must either be kept under control by a handler or prevented from roaming freely on the premises.

It would be wise for employers to take out a personal accident policy to protect everyone employed in connection with the animal house who may be affected by the activity of anti-vivisectionists.

Waste disposal

The most recent Acts controlling the disposal of waste are the Control of Pollution Act 1974, the Control of Pollution (Amendment) Act 1989 and the comprehensive Environmental Protection

Act 1990 (EPA). The HSWA provides for the prevention of emissions of noxious or offensive substances.

It is important for the understanding of the legislation on pollution control to grasp the meaning of certain terms. 'Controlled Waste' is all waste from industrial, commercial and domestic premises. 'Special Waste' is waste which is dangerous to life or has a flash-point of 21°C or below, and contains one or more of the substances listed in the Control of Pollution (Special Waste) Regulations 1980.

The EPA prescribes a 'Duty of Care' for waste management, the purpose of which is to ensure that those dealing with waste take all reasonable steps to avoid pollution of the environment or harm to human health and must take all reasonable measures to:

- prevent the unlawful disposal of waste;
- prevent the escape of waste;
- ensure the waste is only transferred to authorized persons.

The Environmental Protection (Duty of Care) Regulations 1991 established a system of documentation and consignment notes to ensure that waste is adequately described before being transferred and that suitable records are kept for two years.

It is an offence under the EPA to treat, keep or dispose of controlled waste (or knowingly permit controlled waste to be treated, kept or disposed of – so watch where it goes) except under and in accordance with a waste management licence. Within the context of the Waste Management Licensing Regulations 1994, animal by-products are not treated as industrial or commercial waste if collected and transported in accordance with the Animals By-product Order 1992.

The Collection and Disposal of Waste Regulations 1988 allows the storage of controlled waste, pending its disposal elsewhere, on the producer's premises. Where special waste or other hazardous wastes are being stored or handled, COSHH Regulations apply.

The EPA defines statutory nuisance as any of the following:

- any premises;
- smoke emitted from premises;
- any accumulation or deposits;
- any animals;
- noise (including vibrations) emitted from premises

if they are prejudicial to health or are a nuisance (an unlawful interference with the comfort or convenience of another person). The Local Authority now has powers to serve anticipatory abatement notices where it is satisfied that a statutory nuisance is likely to occur. Subject to certain exclusions, it is a defence to prove that the 'best practicable means' were used to prevent or counteract the effects of the nuisance.

The EPA introduced the concept of personal liability for directors and officers of a company where the company is involved in an environmental offence.

Smoke emission

The Clean Air Act 1956 defines smoke as including '... soot, ash, grit and gritty particles emitted in smoke'. The HSWA imposes a duty to use the best practicable means: '... to prevent emission into the atmosphere of noxious or offensive substances, to render harmless or inoffensive such substances as may be emitted'. The Health and Safety (Emissions into the Atmosphere) Regulations 1983 list the 'noxious' and 'offensive' substances and the premises to which the HSWA applies. These lists have been extended by the Health and Safety (Emission into the Atmosphere) Amendment Regulations 1989. These provisions are under the control of the EPA.

HM Inspectors of Pollution (HMIP) may serve an improvement notice on an occupier. If an HMIP thinks the pollution can cause serious personal injury they can serve a prohibition notice.

The Control of Pollution Act 1974 designates powers of investigation to Local Authorities. The Clean Air Acts 1956, 1968 and 1993 permit Local Authorities to establish parts of their district as 'smoke control areas'.

The Control of Smoke Pollution Act 1989 extended the offence of emitting dark smoke beyond the occupier to include any person who causes or permits the emission of dark smoke. 'Dark smoke' is defined as darker than shade 2 on the Ringelmann Scale.

Water pollution

Directive 76/464/EEC and a number of similar Directives deal with the control of the discharge of prescribed substances into water. The details are set out in the Water Resources Act 1991. Discharges of sewage and effluent to watercourses are permitted only if the necessary consent is obtained from the Environment Agency (formerly the National Rivers Authority).

Animal house waste

The COP of the Animals (Scientific Procedures) Act 1986 (Home Office 1989, p7 para 2.33 and p24 para 4.17) recommends: 'A vermin-free collection area should be provided for waste, prior to its disposal. Special arrangements should be made for handling carcasses and radio-active or other hazardous material' and '... infected, toxic or radio-active carcasses must be disposed of so as not to present a hazard'.

Animal carcasses, tissues, body fluids, needles, syringes etc., from animal facilities are classified as 'clinical waste' under the Controlled Waste Regulations 1992 and may also be subject to specific regulations and codes of such bodies as the Advisory Committee on Dangerous Pathogens and the Health Services Advisory Committee.

The safety policy should cover arrangements for labelling, colour-coding, segregation, suitable packaging, storage, transportation and disposal of waste. Records of the disposal of radioactive material must be kept. The hazards associated with cleaning out incinerators should be taken into account, e.g. the presence of sharps and radioactivity. Waste producers are under a duty of care to transfer waste safely to an authorized waste carrier or a waste manager who is licensed to carry out such tasks.

Enforcement (waste disposal)

The Waste Regulations Authorities which may impose specific control are County Councils in most of England, District Councils in Wales, Islands and District Councils in Scotland and specialized authorities elsewhere as in the Metropolitan Areas. The London authority has issued 'Guidelines for the Segregation, Handling and Transport of Clinical Waste (1989)'.

Signs and notices

The Health and Safety (Safety Signs and Signals) Regulations 1996 implement the EC Safety Signs Directive 92/58/EEC. The Regulations require employers to provide and maintain certain safety signs where their risk assessment under Regulation 3 of the Management of Health and Safety at Work Regulations 1992 identifies significant residual risks in respect of which useful information can be provided by the signs.

Safety signs for the purpose of the Regulations include not just conventional signs but also illuminated signs, acoustic and hand signals. Fire signs are also covered. There is a requirement for traffic signs, where necessary, within work places.

In keeping with the Health and Safety Information for Employees Regulations 1989, employers should exhibit the following:

- a thermometer on each floor or in each workroom;
- the poster 'Health and safety law – what you should know';
- a copy or abstract of relevant Regulations;
- a notice specifying the person (or persons) in charge of the first-aid box(es);
- a certificate of insurance as required by the Employers' Liability (Compulsory Insurance) Act 1969;
- any information necessary to comply with a fire certificate;
- local addresses of the relevant enforcing authority, the HSE and the Employment Medical Advisory Service.

Damage limitation

First aid

The duty to provide first aid is dealt with by the Health and Safety (First-Aid) Regulations 1981. The ACOP (COP 42, L74) and guidance were revised and updated in 1997. The regulations lay down three broad duties:

- the duty of the employer to provide first aid;
- the duty of the employer to inform employees of arrangements made in connection with first aid;
- the duty of the self-employed to provide first aid equipment.

First aid is not 'nursing'; it is a specialized and limited activity concerned with damage limitation at the scene of an accident or hazardous occurrence. The Health and Safety (First-Aid) Regulations 1981 do not include the treatment of minor illnesses, such as the administration of tablets and/or medicines, within the definition of first aid.

The 1997 ACOP imposes a duty on employers to carry out assessments to determine the hazards likely or foreseeably present within the particular work place and to provide adequate and suitable first-aid measures accordingly, e.g. the provision of antidotes to toxic hazards if they are present.

The extent and distribution of first-aid facilities and first-aiders should ensure quick and ready access for all employees. The criteria to be considered are: the number of employees, the nature of work activities, the distribution of the workers throughout the establishment and the site location.

First-aid information should be included in an induction programme and all employees should be informed of any changes in arrangements. The first-aid room should be clearly marked and the names and locations of first-aiders should be posted on the door.

The ACOP defines a 'suitable person' for the post of first-aider as someone who holds a first-aid certificate. The certificate is valid for a period of three years. After this time a refresher course and re-examination are required. If there is no first-aider, it is accepted as suitable compliance with the Regulations to appoint a person to take charge of:

- any situation where an ill or injured employee requires attention from any medical practitioner or nurse;
- any first-aid equipment and facilities while the first-aider is absent.

An 'appointed person' should not render any first-aid treatment other than emergency first aid and then only where they have been trained specifically in these procedures, but should be responsible for summoning professional assistance.

If a lone worker sustains a minor injury, he may be able to use a first-aid box, or phone for help. Where more serious injuries are foreseeable, then the absence of a colleague at least to organize help could be construed as insufficient first-aid cover. Intermittent phone calls from outside can be used to confirm a lone worker's status.

Accidents

An accident is an unexpected event causing damage and/or injury, but foresight and planning can reduce the risk of accidents happening and minimize their effects.

The Reporting of Injuries, Diseases and Dangerous Occurrences Regulations 1995 (RIDDOR) imposes duties on persons (not just the employer) responsible for activities at work to report to the enforcing authority (generally the HSE or the Local Authority) the following:

- fatal accidents;

- major injury accidents/conditions;
- dangerous occurrences of a serious nature;
- accidents causing more than three days' incapacity for work;
- certain work-related diseases (refer to the section on industrial diseases);
- matters dealing with the safe supply of gas.

Under the Social Security (Claims and Payments) Regulations 1979, an employee who suffers personal injury through an accident at work must inform the employer immediately, either orally or in writing, or as soon as possible. Employers must take reasonable steps to investigate the circumstances of every accident notified to them, and any discrepancy between their findings and the information provided by the employee must be recorded. The management's established safety policy should explain the internal reporting procedures. A guide to RIDDOR is available from the HSE (HSE 31 1995, L73 1996). Not only accidents but also such occurrences as the escape of pathogens are of significance and must be noted. 'Near misses' may indicate the need to revise safety policy.

Record-keeping

The following details of any accident or dangerous occurrence must be kept and retained for three years: date and time; name and occupation of the victim; details of the injury; place; circumstances. In the case of a reportable disease the date of diagnosis, the nature of the disease and occupation of the person affected should be recorded.

An accident book is required under the Social Security (Claims and Payments) Regulations 1979 for premises where 10 or more persons are employed. Even minor injuries must be recorded with the following details: name, address and occupation of the injured person; date and time of accident; place; cause and nature of injury; name, address and occupation of person making the entry.

It is an offence to make a false entry in relation to these records and reports.

Liability

This is a complicated area of law associated with tort – that is, a civil offence as opposed to a criminal offence. A civil case involves a claim for compensation rather than an imposition of a penalty. The person making the claim for damages (the plaintiff) must establish that what he asserts is true. The other party (the defendant) may bring evidence to refute this assertion. In a civil case it is enough that the claimant proves his case on 'a balance of probabilities'.

Occupiers' liability

Under the Occupiers' Liability Act 1984, an occupier owes 'a duty of care' to his visitors. The HSWA reinforces this demand. The information given to them should be commensurate with their involvement in the activities taking place on the premises and the nature and the extent of the hazards they may encounter. These instructions and warnings are best given in printed form on a card.

Trespassers. An occupier who knows that trespassers may encounter hazards, owes them the 'humanity duty' – a duty to take such steps as common sense or common humanity would dictate to prevent harm.

Vicarious liability

An employer may be held civilly or criminally liable for the negligent or unlawful acts of an employee, even though it can be shown that the employee wilfully disobeyed the express instructions of his employer. Employees may incur liability under the HSWA for not taking reasonable care of themselves and others in the work place or for being reckless in regard to equipment.

Liability may be incurred for the actions of contractors working under contract for an employer. The work of contractors, therefore, should be closely controlled.

Employers' liability

Since the Law Reform (Personal Injuries) Act 1948, an employer is liable for injuries to employees caused by the negligence of another employee. The Employers' Liability (Defective Equipment) Act 1969 makes an employer liable for injury sustained by an employee where it is caused by a defect in equipment attributable wholly or partly to the fault of a third party.

The strictness of employers' liability cannot be over-stressed. Unless there is continuing and determined action by management to enforce its own rules, it is likely that the employer will be held at least partially liable for any injury incurred.

Dismissals can occur for health and safety reasons, for example:

- the employee's unsafe or reckless behaviour;
- the employee's inability to do a job safely;
- the employee's refusal to obey an order if it is a safety rule which he will not obey.

The grounds for any dismissal must be reasonable. A case of constructive and unfair dismissal may arise if an employee's resignation is brought about by the employer's unsafe practices

Claims and benefits

Compensation is a payment to make amends for loss or injury to a person or property. In health and safety there are two basic types of compensation to be considered:

a) Compensation in respect of accidents arising out of or in the course of a person's employment, or in respect of disease contracted in the course of employment, may be available in the form of Statutory Sick Pay (SSP), National Insurance sickness benefit or NI disablement benefit.
b) Damages for loss suffered owing to negligence, breach of statutory duty or other tort which must be proved and claimed for in the Civil Court.

An employee who suffers an industrial injury and who is eligible for SSP will receive payment at the prescribed rate from the employer under this scheme. If the employee is excluded from receiving SSP, the employer must provide a change-over form SSP.

An accident which happens when an employee is at work is treated as having occurred out of and in the course of employment, unless there is evidence to the contrary. If employees are injured whilst travelling to or from the place of work in transport provided by the employer, they will be treated as having been at work at the time.

Benefits are also payable if an employee is injured whilst doing something expressly prohibited by the employer or in contravention of any of the 'relevant statutory provisions', provided that what the employee was doing was for the purpose of the employer's business and was within the scope of the employee's job. Likewise, an employee who is injured while going to the assistance of another person who has been injured or is in danger of being injured, or while seeking to avert or minimize serious damage to property, will also qualify to be paid benefit.

An employee who is injured through another person's misconduct or negligence, or by the behaviour of an animal, or by being struck by lightning or any object, will be entitled to receive benefit, provided the injuries were sustained in the course of the employment and the employee did not directly or indirectly induce or contribute to the happening of the accident.

Industrial diseases

If, in the opinion of the examining doctor, a person may not continue to work on a particular process without serious risk to their health, the doctor may suspend that person for a specified or indefinite period of time.

The Employment Protection (Consolidation) Act 1978, which consolidates the 1975 Act, provides that an employee who is suspended on medical grounds in consequence of any requirement imposed under statutory provision is entitled to be paid remuneration by the employer, while he is suspended, for a maximum of 26 weeks. Such employees will not be entitled to be paid if they do not comply with any reasonable requirements imposed by the employer or refuse to accept an offer of suitable alternative work made by the employer during the period of suspension. An employee needs to be employed for at least a month and for more than 16 hours each week to benefit from these provisions. A person dismissed on health grounds may bring a complaint of unfair dismissal to an industrial tribunal.

It is important to realize that in respect of these various claims, completing or signing reports does not constitute an admission of liability.

Bases of claims

Breach of statutory duty

A person who fails to comply with statutory requirements may also be liable in civil law for injury or damage which is the consequence of that failure.

Negligence

This is the basis of most claims in the Civil Court. Liability arises in negligence when it is reasonably foreseeable that some harm will be caused and the failure to prevent it causes damage, injury or death.

Contributory negligence. This may apply where a person suffers damage as a result partly of his own fault and partly of the fault of any other person. Contributory negligence can be established by proof of lack of care by the plaintiff for his own safety. It is operative where the plaintiff's own carelessness does not actually contribute to the cause of the injury but merely increases the extent of it. This factor would be germane in cases where protective equipment was not being used properly by the injured worker.

Strict liability

'The person who for his own purposes brings on his land and collects and keeps there anything likely to do mischief if it escapes must keep it in at his peril and, if he does not do so, is *prima facie* answerable for all the damage which is the natural consequences of its escape.' Escaping animals heading across busy roads or pathogens floating in the air may be pertinent to this type of liability.

Nuisance

Some nuisances are expressly forbidden by laws such as the Clean Air Act 1993. Smoke, smells, noise and vibration may all be actionable in nuisance. The Control of Pollution Act 1974 empowers Local Authorities to serve notice on an occupier requiring abatement of a nuisance.

Liability for animals

The Animals Act 1971 consolidated much of the old complex Common Law on responsibility for injuries and damage caused by animals. The Act makes the keeper of a non-domesticated species liable for any harm it causes. In the case of domesticated animals the keeper is liable for damage which is due to some unusual propensity of the animal of which the keeper, or those acting on his behalf, are aware. S.6(5) has special relevance for animal staff: 'Where a person employed as a servant by a keeper of an animal incurs a risk incidental to his employment he shall not be treated as accepting it voluntarily'. This means that a salient defence in tort against liability, *voluntas non fit injuria*, is not available to an employer if his employee, injured by an animal in his charge, sues for compensation.

Protection by insurance, etc.

It is normal to ensure that potential liabilities are insured against. It is important to consider the sum insured, the risk involved, the person covered and whether legal costs are included in the policy. Every employer is obliged under the Employer's Liability (Compulsory Insurance) Act 1969 to insure against liability for bodily injury or disease sustained by his employees. The Employers' Liability (Compulsory Insurance) Regulations 1998 oblige an employer to insure for a minimum of £5 million in respect of claims arising out of any one occurrence. A copy of the certificate of insurance must be displayed at the place of business.

Where to find the law

United Kingdom

Statutes and Statutory Instruments can be obtained from:
The Stationery Office (HMSO), PO Box 276, London SW8 5DT.
The following publications are updated regularly:
Halsbury's Laws of England. London: Butterworth
Halsbury's Statutory Instruments. London: Butterworth

European legislation

Information on this subject can be obtained from:
Jean Munnat House, 8 Storey's Gate, London SW18 3AT. Tel: 020 7973 1900
Office for Official Publications of the EU, 2 Rue Mercier L-2985, Luxembourg
Celex (Communitas Europae Lex) available on CD-ROM, covers a compendium of European Legislation
ECLAS (European Commission's Library Automated System). Eurobases, Commission of the EU, Rue de Loi 200, B-1049 Brussels, Belgium

UK Codes of Practice etc.

Information on this legislative material and official literature can be obtained from:
Health and Safety Executive, Information Centre, Broad Lane, Sheffield S3 7HQ. Infoline: 0541 545500. Fax: 0114 289 2333.
HSE Books, PO Box 1999, Sudbury, Suffolk CO10 6FS. Tel: 01787 881165.
The Stationery Office Ltd, PO Box 276, Nine Elms Lane, London SW8 5DT. Tel: 020 7873 0011.

Information on British Standards (some incorporate European standards) is available from:
British Standards Institution, Standards Sales – Customer Services, 389 Chiswick High Road, London W4 4AL. Information Centre Tel: 020 8996 7111.

European Standards are produced by:
Comité Européen de Normilisation, Central Secretariat, Rue de Stassart 36, B-1050 Brussels, Belgium

Other sources of UK/EC information are databases such as:
Barbour Index, Health and Safety Professional Database, Barbour Index plc, Windsor, Berkshire.
Technical Indexes Ltd, Willoughby Road, Bracknell, Berkshire RG12 8DW.

NB. Any specific problems involving litigation should be referred as soon as possible to professional legal experts.

The publications listed below are quoted in the text or are other sources of relevant information.

Statutes

Agriculture (Miscellaneous Provisions) Act 1968
Animals Act 1971
Animal Health Act 1981
Animals (Scientific Procedures) Act 1986
Clean Air Act 1956, 1968, 1993
Consumer Protection Act 1987
Control of Pollution Act 1974
Control of Pollution (Amendment) Act 1989
Control of Smoke Pollution Act 1989
Crown Proceedings Act 1947
Dangerous Wild Animals Act 1976
Employers' Liability (Compulsory Insurance) Act 1969
Employers' Liability (Defective Equipment) Act 1969
Employment Protection (Consolidation) Act 1978
Environmental Protection Act 1990
European Communities Act 1972
Factories Act 1961
Fire Precautions Act 1971
Guard Dogs Act 1975
Health and Safety at Work etc. Act 1974
Industrial Diseases (Notification) Act 1981
Law Reform (Personal Injuries) Act 1948
Medicines Act 1968
Misuse of Drugs Act 1971
National Health Service (Amendment) Act 1986
Occupiers' Liability Act 1984
Offices, Shops and Railway Premises Act 1963
Poisons Act 1972
Public Health Act 1936 and 1961
Radioactive Substances Act 1948, 1960, 1993
Water Resources Act 1991

Statutory Instruments (SI): Regulations, Orders and Rules

Many statutory instruments will have been amended since enactment

Animals and Animal Products (Import and Export) Regulations 1993 (SI 3247), 1995 (SI 2428)
Animals By-Product Order 1992 (SI 3303)
Anthrax Order 1991 (SI 2814)
Chemicals (Hazard Information and Packaging for Supply) Regulations 1994 (SI 3247) (CHIP 2), various amendments to 1999 (SI 197)
Collection and Disposal of Waste Regulations 1988 (SI 819)
Control of Pollution (Special Waste) Regulations 1980 (SI 1709)
Controlled Waste Regulations 1992 (SI 588)
COSHH Regulations 1988 (SI 1657), 1994 (SI 3246) and 1999 (SI 437)
Electricity at Work Regulations 1989 (SI 635)
Employers' Health and Safety Policy Statements (Exception) Regulations 1975 (SI 1584)
Employers' Liability (Compulsory Insurance) Regulations 1998 (SI 2573)
Environmental Protection (Duty of Care) Regulations 1991 (SI 2839)
Environmental Protection (Prescribed Processes and Substances) Regulations 1991 (SI 472)
Fire Precautions (Application for Certificate) Regulations 1989 (SI 77)
Fire Precautions (Workplace) Regulations 1997 (SI 1840)
Genetically Modified Organisms (Contained Use) Regulations 1992 (SI 3217) as amended 1998 (SI 1548)
Genetically Modified Organisms (Deliberate Release) Regulations 1992 (SI 3280), 1995 (SI 304)
Genetically Modified Organisms (Risk Assessment) (Records and Exemptions) Regulations 1996 (SI 1106)
Health and Safety (Display Screen Equipment) Regulations 1992 (SI 2792)
Health and Safety (Emissions into the Atmosphere) Regulations 1983 (SI 943), amended 1989 (SI 319)
Health and Safety (Enforcing Authority) Regulations 1998 (SI 494)
Health and Safety (First-Aid) Regulations 1981 (SI 917)
Health and Safety Information for Employees Regulations 1989 (SI 682)
Health and Safety Policy Statement (Exception) Regulations 1975 (SI 1584)
Health and Safety (Safety Signs and Signals) Regulations 1996 (SI 341)
Health and Safety (Training for Employment) Regulations 1990 (SI 1380)
Health and Safety (Young Persons) Regulations 1997 (SI 135)
Ionising Radiations Regulations 1985 (SI 1333)
Ionising Radiations (Outside Workers) Regulations 1993 (SI 2379)
Lifting Plant and Equipment (Records of Test and Examination, etc.) Regulations 1992 (SI 195)
Management of Health and Safety at Work Regulations 1992 (SI 2051), as amended 1994 (SI 2865) etc.
Manual Handling Operations Regulations 1992 (SI 2793)
Misuse of Drugs (Safe Custody) Regulations 1973 (SI 798)
Misuse of Drugs Regulations 1985 (SI 2066), 1995 (SI 3244), 1996 (SI 1597)
Noise at Work Regulations 1989 (SI 1790)
Notification of New Substances Regulations 1993 (SI 3050) (known as NONS)

Personal Protective Equipment at Work Regulations 1992 (SI 2966)
Personal Protective Equipment (EC Directive) Regulations 1992 (SI 3139) amended 1994 (SI 2326) and 1996 (SI 3039)
Poisons Rules 1982 (SI 218)
Poisons List Order 1978 (SI 2)
Prescription Only Medicines (Human Use) Order 1997 (SI 1830)
Pressure Systems and Transportable Gas Containers Regulations 1989 (SI 2169)
Provision and Use of Work Equipment Regulations 1992 (SI 2932), 1998 (SI 2306)
Public Information for Radiation Emergencies Regulations 1992 (SI 2997)
Rabies (Importation of Dogs, Cats and other Mammals) Order 1974 (SI 2211)
Rabies Virus Order 1979 (SI 135)
Radioactive Substances (Substances of Low Activity) Exemption (Amendment Order) 1992 (SI 647)
Reporting of Injuries, Diseases and Dangerous Occurrences Regulations 1995 (SI 3163) (known as RIDDOR)
Safety Representatives and Safety Committees Regulations 1977 (SI 500)
Social Security (Claims and Payments) Regulations 1979 (SI 628)
Social Security (Industrial Injuries) (Prescribed Diseases) Regulations 1985 (SI 967)
Special Waste Regulations 1996 (SI 972, 2019)
Special Waste (Amendment) Regulations 1996 (SI 2019), 1997 (SI 251)
Supply of Machinery (Safety) Regulations 1992 (SI 3073)
Waste Management Licensing Regulations 1994 (SI 1056)
Waste Management Regulations 1996 (SI 634)
Working Time Regulations 1998 (SI 1833)
Workplace (Health, Safety and Welfare) Regulations 1992 (SI 3004)
Zoonoses Order 1975 (SI 1030)

European Directives

Published in the *Official Journal of the European Communities*

67/548/EEC: on the approximation of laws, regulations and administrative provisions relating to the classification, packaging and labelling of dangerous substances (Dangerous substances directive)

76/464/EEC: on pollution caused by certain dangerous substances discharged into the aquatic environment of the community

80/1107/EEC: on the protection of workers from the risks related to exposure to chemical, physical and biological agents at work

86/188/EEC: exposure to noise at work – protection of workers from the risks related to exposure to noise at work

86/609/EEC: on the approximation of laws, regulations and administrative provisions of member states regarding the protection of animals used for experimental and other scientific purposes

89/391/EEC: on the introduction of measures to encourage improvements in the safety and health of workers at work (general principles on health and safety – Framework directive)

89/392/EEC (amended in 91/368/EEC) and 93/37/EC: on the approximation of the laws of Member States relating to machinery (Machinery directives)

89/654/EEC: concerning the minimum safety and health requirements for the work place (Workplace directive)

89/655/EEC: concerning the minimum safety and health requirements for the use of work equipment by workers at work (Use of work equipment directive). Amended 95/63/EC

89/656/EEC: on the minimum health and safety requirements for the use by workers of personal protective equipment at the work place (Personal protective equipment directive)

89/686/EEC: on the approximation of the laws of the Member States relating to personal protective equipment (Personal protective equipment (PPE) directive); see also 93/68/EEC, 93/95/EEC, 96/58/EEC

90/219/EEC: on the contained use of genetically modified microorganisms. Amended 98/81/EC

90/220/EEC: on the deliberate release into the environment of genetically modified organisms

90/269/EEC: on the minimum health and safety requirements for the manual handling of loads where there is a risk particularly of back injury to workers (Manual handling of loads directive)

90/270/EEC: on the minimum safety and health requirements for work with display screen equipment (Display screen equipment directive)

90/394/EEC: on the protection of workers from the risks related to exposure to carcinogens at work (Carcinogens directive); since amended 97/42/EEC, 98/C 123/12

90/425/EEC and 92/60/EEC: veterinary and zootechnical checks applicable in intra-community trade in certain live animals and products with a view to the completion of the internal market

90/641/Euratom: on the operational protection of outside workers exposed to the risks of ionizing radiation during their activities in controlled areas

90/679/EEC (amended 93/88/EEC): on the protection of workers from risks related to exposure to biological agents at work (Biological agents directive)

91/322/EEC: on establishing indicative limit values by implementing Council Directive

91/383/EEC: supplementing the measures to encourage improvements in the safety and health at work of workers with a fixed duration employment relationship or a temporary employment relationship (Temporary workers directive)

92/58/EEC: on the minimum requirements for the provision of safety and/or health signs at work (Safety signs directive)

92/65/EEC: laying down animal health requirements governing trade in and imports into the Community of animals, semen, ova, and embyos not subject to animal health requirements laid down in specific Community rules referred to in 90/425/EEC (BALAI agreement)

92/85/EEC: on the introduction of measures to encourage improvements in the safety and health at work of pregnant workers etc. (Pregnant workers directive)

93/67/EEC: risks to man and the environment of substances notified in accordance with Council Directive 67/548/EEC (Risk assessment directive)

93/88/EEC: providing a community classification of biological agents

93/104/EEC: concerning certain aspects of the organisation of working time (Working time directive)

94/33/EC: on the protection of young people at work (Young workers directive)

96/29/Euratom: laying down the basic safety standards for the protection of the health of workers and the general public and workers against the dangers arising from ionizing radiation

96/55/EC: on the approximation of laws, regulations and administrative provisions of the Member States relating to restrictions on the marketing and use of certain dangerous substances and preparations which relates to supply of specified substances for use at work.

98/24/EC: protection of health and safety of workers from risks related to chemical agents at work

Codes, Guidance Notes and advisory literature

These publications are obtainable from the HSE or The Stationery Office at the addresses given above.

The Guidance Note series is divided into five areas:
Chemical Safety (CS)
Environmental Hygiene (EH)
General Series (GS)
Medical Series (MS)
Plant and Machinery (PM)

Other abbreviations used in reference to safety literature are:

COP	Codes of Practice
EH	Environmental Hygiene Series
HSC	Health and Safety Commission Leaflets
HS(G)	Health and Safety: Guidance Booklets
HS(R)	Health and Safety: Regulations Booklets
IND(G)	Industrial General Leaflets
L	Legislation Series (Leaflets)
SIR	Specialist Inspector Reports

There is an abundance of this literature and much of it can be obtained free of charge. It is updated as the need arises and the numbered, abbreviated reference is usually sufficient identification. The following lists, grouped according to topics in the order in the text, include the more relevant publications which may reflect in some way on legal obligations.

Health and Safety at Work etc. Act (HSWA)

HSC 2	Health and Safety at Work, etc. Act. The Act outlined 1994
HSC 3	Advice to employers 1995

HSC 5	Health and Safety at Work, etc. Act. Advice to Employees 1994
HSC 6	Writing a safety policy statement: Advice to employers, 1990
HSC	Management of health and safety at work regulations. ACOP 1992
HSE 23(Rev)	Health and safety legislation and trainees (A guide for employers) 1991
HS(G) 48	Human factors in industrial safety 1989
HS(G) 61	Surveillance of people exposed to health risks at work 1990
HS(G) 65	Successful health and safety management 1997
IND(G) 76L	Safe systems of work 1992
IND(G) 275	Managing health and safety – Five steps to success 1998

Inhalation

HS(G) 37	An introduction to local exhaust ventilation 1993
HS(G) 53	The selection use and maintenance of respiratory protective equipment 1998
HS(G) 54	The maintenance, examination and testing of local exhaust ventilation 1998

Noise

HS(G) 56	Noise at work: Noise assessment information and control 1990
IND(G) 99(rev)	Noise at work – A guide for employees 1998

Light

HS(G) 38	Lighting at work 1997

Lifting

HS(R)26	Guidance on the legal and administrative measures taken to implement the European Community Directives on Lifting and Mechanical Handling Appliances and Electrically Operated Lifts 1987
IND(G) 143L	Getting to grips with manual handling; a short guide for employers, 1993
IND(G) 110(rev)	Lighten the load: Guidance for employees on musculoskeletal disorders 1993
L20	A guide to the Lifting Plant and Equipment (Records of Testing and Examinations etc.) Regulations 1992
L23	Manual handling: guidance on regulations 1992

Hazardous substances

EH 40	Occupational exposure limits (annual)
EH 64	Summary criteria for occupational exposure limits 1998
HSC	Safety in health service laboratories: safe working and the prevention of infection in clinical laboratories 1991
HSE	The occupational zoonoses 1993
HSE 14 MDHS	Methods for the determination of hazardous substances (various years)
HS(G) 97	A step by step guide to COSHH assessments 1993
IND(G)91L(rev)	Drug abuse at work: a guide to employers 1990
IND(G)136(rev1)	COSHH: A brief guide to the regulations 1999
IND(G)181(L)	The complete idiot's guide to CHIP 2 1995
L5	General COSHH ACOP, Carcinogens ACOP (Control of Carcinogenic Substances) and Biological Agents ACOP (Control of Biological Agents) 1999 Control of Substances Hazardous to Health Regulations 1999
L62	ACOP Safety data sheets for substances and preparations dangerous for supply 1995 (ref CHIP2)
L63	Approved guide to the classification and labelling of substances and preparations dangerous for supply 1993 (ref CHIP2)
L86	Control of substances hazardous to health in fumigation operations 1996 (ACOP)
L115	Approved supply list. Information approved for the classification and labelling of substances dangerous for supply. 1998 (ref CHIP2)

Radiation

HS(G) 49	The examination and testing of portable radiation instruments for external radiations 1990
HS(G) 90	VDUs. An easy guide to the regulations 1997
IND(G)36(revl)	Working with VDUs 1998
L7	Fees and charges Radioactive Substances Act Regulations 1993, 1995
L26	Display screen equipment work. Guidance on regulations 1992

Equipment

IACL16(Rev)	Protective clothing and footwear 1991
HSC	COP 37 Safety of pressure systems, 1990
HS(G) 57	Seating at work 1991, 1997
HS(R) 30	A guide to the Pressure Systems and Transportable Gas Containers Regulations 1990
L22	ACOP. Provision and use of work equipment 1998
L25	Personal protective equipment at work regulations. Guidance on regulations 1992

Electricity

HS(R) 25	A memorandum of guidance on the Electricity at Work Regulations 1989

Other topics

Home Office	Fire precautions in the workplace. Information for employers about the Fire Precautions (Workplace) Regulations 1997, 1997
HSE	Watch your step: Prevention of slipping, tripping and falling accidents at work 1985
HSE 4(rev)	Employers' Liability (Compulsory Insurance) Act 1969: a short guide 1993
HSE 31	Everyone's guide to RIDDOR 95 1995
HS(G) 51	The storage of flammable liquids in containers 1990, 1998
IND(G) 113L	Your firm's injury records and how to use them 1991
IND(G) 118(L)	Policy statement on publication of reports on incidents 1991
L24	Workplace health, safety and welfare regulations 1992. ACOP and guidance 1992
L73	RIDDOR 95. A guide to the reporting of injuries, diseases and dangerous occurrences regulations 1995. 1996
L74	First-aid at work: The health and safety (first-aid) regulations 1981 approved code of practice and guidance 1997

British Standards

EN is Norme Européene or European Standard. This list is a small selection of BS publications.

Inhalation, noise, sight

BS EN 141	Specifications for gas filters and combined filters used in respiratory protective equipment 1991
BS EN 143	Specification for particle filters used in respiratory protective equipment 1991
BS EN 149	Filtering half masks to protect against particles 1992
BS EN 165/166	Personal eye protection 1996
BS EN 352	Hearing protectors – safety requirements and testing 1993–97
BS EN 458	Hearing protectors – recommendations for selection, use, care and maintenance. Guidance document 1994

Radiation

BS 5288	Sealed radioactive sources 1976
BS 6090	Personal photographic dosemeters 1981

Computer use

BS 7033	Guide to the accommodation and operating environment for information technology (IT) equipment 1996
BS 7179 pt 5/6	Ergonomics of design and use of visual display terminals (VDTs) in offices 1990
BS EN 60950	Information technology equipment including electrical business equipment 1992

Equipment

BS 2646	Autoclaves for sterilisation in laboratories, 1991
BS 3813	Incinerators for waste from trade and residential premises
BS 5378	Safety signs and colours 1980/82
BS 5426	Workwear and career wear 1993
BS 5726 pt 4	Microbiological safety cabinets. Recomendations for selection, use and maintenance 1992
BS 7258	Laboratory fume cupboards 1994
BS 7320	Sharps containers 1990

Fire

BS 5266	Emergency lighting 1988
BS 5445	Components of automatic fire detection systems 1977-84
BS 5499	Fire safety signs, notices and graphic symbols 1990
BS 5839	Fire detection and alarm systems in buildings, 1988
BS EN 3 (1-6)	Portable fire extinguishers 1996

Other references and suggested additional reading

Advisory Committee on Dangerous Pathogens (ACDP) (1997) *Working safely with research animals: management of infection risks*. Sudbury: HSE Books

Advisory Committee on Dangerous Pathogens (ACDP) (1998) *Working safely with simians: Management of infection risks*. Sudbury: HSE Books

Association of the British Pharmaceutical Industry (1985) *Guidelines for the control of occupational exposure to therapeutic substances*. London: ABPI

Bell JC, Palmer SR, Payne JM (1988) *Zoonoses: infections transmitted from animals to man*. London: Edward Arnold

British Veterinary Association (1991) *COSHH: BVA guide to the initial assessment in veterinary practice*. London: BVA Publications

Collins CH, ed (1985) *Safety in biological laboratories*. London: Institute of Biology

Cooper ME (1987) *An introduction to animal law*. London: Academic Press

Croner Publications Ltd (Kingston upon Thames, KT2 6SR) (updated regularly) such as:

Croner's Dangerous Substances

Croner's Health and Safety Case Law Index

Croner's Substances Hazardous to Health
Croner's Premises Management
Croner's Environmental Management
Croner's Health and Safety at Work

Dewis M (1998) *Tolley's Health and Safety at Work*. Croydon: Tolley Publishing Company Ltd (regularly updated)

Health and Safety Commission (1995) Advisory Committee on Dangerous Pathogens. *Categorisation of biological agents according to hazard categories of containment.* 4th edn. Sudbury: HSE Books

Home Office (1989) *Code of practice for the housing and care of animals used in scientific procedures*. London: HMSO

James I, Preece D (1997) *Jordan's Health and Safety Management 1998*. Bristol: Jordan Publishing Ltd